SIGNIFICANT CHANGES TO THE

INTERNATIONAL RESIDENTIAL CODE®

2021 EDITION

SIGNIFICANT CHANGES TO THE
International Residential Code®

2021 EDITION

International Code Council

ICC Staff:

Executive Vice President and Director of Business Development:
  Mark A. Johnson

Senior Vice President, Business and Product Development:
  Hamid Naderi

Vice President and Technical Director, Products and Services:
  Doug Thornburg

Senior Marketing Specialist:
  Dianna Hallmark

ISBN: 978-1-952468-19-3 (print)
ISBN: 978-1-955052-44-3 (PDF download)

Project Head:	Doug Thornburg
Publications Manager:	Anne F. Kerr
Cover Design:	Ricky Razo

COPYRIGHT © 2021
by INTERNATIONAL CODE COUNCIL, INC.

ALL RIGHTS RESERVED.

This publication is a copyrighted work owned by the International Code Council, Inc. (ICC). Without advance written permission from the ICC, no part of this publication may be reproduced, distributed or transmitted in any form or by any means, including, without limitation, electronic, optical or mechanical means (by way of example, and not limitation, photocopying or recording by or in an information storage retrieval system). For information on use rights and permissions, please contact: ICC Publications, 4051 Flossmoor Road, Country Club Hills, IL 60478; phone: 1-888-ICC-SAFE (422-7233).

The information contained in this document is believed to be accurate; however, it is being provided for informational purposes only and is intended for use only as a guide. Publication of this document by the ICC should not be construed as the ICC engaging in or rendering engineering, legal or other professional services. Use of the information contained in this book should not be considered by the user to be a substitute for the advice of a registered professional engineer, attorney or other professional. If such advice is required, it should be sought through the services of a registered professional engineer, licensed attorney or other professional.

Trademarks: "International Code Council," the "International Code Council" logo, "ICC," the "ICC" logo, "International Residential Code," "IRC," "International Building Code," "IBC," "International Plumbing Code," "IPC," "International Mechanical Code," "IMC," "International Fire Code," "IFC", "International Existing Building Code," "IEBC," "International Code Council Performance Code," "ICCPC" and other names and trademarks appearing in this book are registered trademarks of the International Code Council, Inc., and/or its licensors (as applicable), and may not be used without permission.

Errata on various ICC publications may be available at www.iccsafe.org/errata.

First Printing: March 2021

PRINTED IN THE USA

Contents

PART 1
Administration
Chapters 1 and 2 1

- **R102.7.1**
  Additions, Alterations or Repairs 2

- **R202**
  Definition of Emergency Escape and Rescue Opening 4

- **R202**
  Definition of Townhouse 6

PART 2
Building Planning
Chapter 3 7

- **R301.1.4**
  Intermodal Shipping Containers 9

- **R301.2**
  Wind Speeds 11

- **Table R301.2.1(1)**
  Component and Cladding Wind Pressures 13

- **R301.2.1.1**
  Special Wind Regions 16

- **R301.2.2.6**
  Irregular Buildings in Seismic Areas 18

- **R301.2.2.10**
  Seismic Anchorage of Water Heaters 22

- **R301.3**
  Story Height 23

- **R302.2**
  Townhouses 26

- **R302.3**
  Two-Family Dwelling Separation 29

- **R302.4**
  Dwelling Unit Rated Penetrations 31

- **R302.5**
  Dwelling-Garage Opening Protection 33

- **R303.1**
  Mechanical Ventilation 35

- **R305.1**
  Ceiling Height 37

- **R308.4.5**
  Glazing and Wet Surfaces 39

- **R308.6**
  Skylight Glass Retention Screens 41

- **R310.1**
  Emergency Escape and Rescue Opening Required 44

- **R310.2**
  Emergency Escape and Rescue Openings 46

- **R310.3, R310.4**
  Area Wells for Emergency Escape and Rescue Openings 49

iii

CONTENTS

- **R310.5, R310.6, R310.7**
  Emergency Escape and Rescue Openings in Existing Buildings ... 53
- **R311.7, R311.8**
  Stairways and Ramps ... 56
- **R311.7.7**
  Stairway and Landing Walking Surface ... 58
- **R312.2**
  Window Fall Protection ... 59
- **R314.3**
  Smoke Alarm Locations ... 61
- **R315.2.2**
  Carbon Monoxide Alarms ... 63
- **R317.1**
  Protection of Wood Against Decay ... 65
- **R320**
  Accessibility ... 69
- **R323**
  Storm Shelters ... 71
- **R324.3**
  Photovoltaic Systems ... 73
- **R326**
  Habitable Attics ... 76

PART 3
Building Construction
Chapters 4 through 10 ... 80

- **Table R403.1(1)**
  Footings Below Light-Frame Construction ... 82
- **R406.2**
  Foundation Waterproofing ... 87
- **R408.8**
  Vapor Retarder in Crawlspaces ... 89
- **R506.2.3**
  Vapor Retarders Under Concrete Slabs ... 91
- **R507**
  Deck Loads ... 93
- **R507.3**
  Deck Footings ... 95
- **R507.4**
  Deck Posts ... 97
- **R507.5**
  Deck Beams ... 100
- **R507.6**
  Deck Joists ... 105
- **R507.7**
  Decking ... 108
- **R507.10**
  Exterior Guards ... 110
- **Table R602.3(1)**
  Fasteners – Roof and Wall ... 112
- **Table R602.3(1)**
  Fasteners – Roof Sheathing ... 114
- **Table R602.3(2)**
  Alternate Attachments ... 116
- **R602.9**
  Cripple Walls ... 118
- **R602.10.1.2**
  Location of Braced Wall Lines ... 120
- **R602.10.2.2**
  Location of Braced Wall Panels (BWPs) ... 123
- **R602.10.3(1)**
  Bracing for Winds ... 125
- **Table R602.10.3(3)**
  Seismic Wall Bracing ... 127
- **Table R602.10.3(4)**
  Adjustment Factors – Seismic ... 129
- **R602.10.6.5**
  Stone and Masonry Veneer ... 132
- **R609.4.1**
  Garage Doors ... 135
- **R702.7**
  Vapor Retarders ... 137
- **R703.2, R703.7.3**
  Water-Resistive Barriers ... 141
- **Table R703.8.4(1)**
  Veneer Attachment ... 144
- **R703.11.2**
  Vinyl Siding Installation Over Foam Plastic Sheathing ... 146
- **R704**
  Soffits ... 148
- **R802**
  Wood Roof Framing ... 153
- **Table R802.5.2(1)**
  Heel Joint Connections ... 155

- **R802.6**
  Rafter and Ceiling Joist Bearing   158

- **Table R804.3**
  CFS Roof Framing Fasteners   160

- **R905.4.4.1**
  Metal Roof Shingle Wind Resistance   163

PART 4
Energy Conservation
Chapter 11   165

- **N1101.6**
  Definition of High-Efficacy Light Sources   167

- **N1101.7**
  Climate Zones   168

- **N1101.13**
  Compliance Options   171

- **N1101.13.5**
  Additional Energy Efficiency Requirements   174

- **N1101.14**
  Permanent Energy Certificate   176

- **N1102.1**
  Building Thermal Envelope   178

- **Tables N1102.1.2 and N1102.1.3**
  Insulation and Fenestration Requirements   180

- **N1102.2**
  Ceiling Insulation   184

- **N1102.2.4**
  Access Hatches and Doors   187

- **N1102.2.7**
  R-Value Reduction for Walls with Partial Structural Sheathing   189

- **N1102.2.7**
  Floor Insulation   191

- **N1102.2.8**
  Unconditioned Basement   193

- **N1102.4 and Table N1102.4.1.1**
  Building Air Leakage and Testing   195

- **N1102.4.6**
  Air-Sealed Electrical Boxes   200

- **N1103.3**
  Duct Installation   202

- **N1103.3.5**
  Duct Testing   205

- **N1103.6**
  Mechanical Ventilation   207

- **N1104**
  Lighting Equipment   210

- **N1105 and Table N1105.2**
  Total Building Performance Analysis   212

- **N1106 and Table N1106.2**
  Energy Rating Index Analysis   216

- **N1108**
  Additional Efficiency Package Options   220

PART 5
Mechanical
Chapters 12 through 23   223

- **M1505**
  Balanced Ventilation System Credit   224

- **M1802.4**
  Blocked Vent Switch for Oil-fired Appliances   227

- **M2101**
  Hydronic Piping Systems Installation   228

PART 6
Fuel Gas
Chapter 24   231

- **G2403**
  Definitions of Point of Delivery and Service Meter Assembly   232

- **G2414.8.3**
  Threaded Joint Sealing   234

- **G2415.5**
  Fittings in Concealed Locations   236

- **G2427.5.5.1**
  Chimney Lining   237

- **G2427.8**
  Through-the-wall Vent Terminal Clearances   239

- **G2439.5**
  Makeup Air for Dryer Installed in a Closet   242

- G2447.2
  Commercial Cooking Appliances Prohibited 243

PART 7
Plumbing
Chapters 25 through 33 **245**

- P2503.5.1
  Drain, Waste and Vent Systems Testing 246

- P2708.4, P2713.3
  Shower and Bathtub Control Valves 248

- P2904
  Installation Practices for Residential Sprinklers 250

- P2905.3
  Length of Hot Water Piping to Fixtures 254

- P3005.2.10.1
  Removable Fixture Traps as Cleanouts 255

- P3011
  Relining of Building Sewers and Building Drains 256

PART 8
Electrical
Chapters 34 through 43 **259**

- E3601.8
  Emergency Service Disconnects 261

- E3606.5
  Service Surge-Protective Device 263

- E3703.4
  Bathroom Branch Circuits 264

- E3703.5
  Garage Branch Circuits 266

- E3901.4
  Kitchen Countertop and Work Surface Receptacles 268

- E3902
  GFCI Protection for 250-Volt Receptacles 271

- E3902.5
  GFCI Protection for Basement Receptacles 274

- E3902.10
  GFCI Protection for Indoor Damp and Wet Locations 276

PART 9
Appendices
Appendix A through W **277**

- AF104
  Radon Testing 278

- Appendix AU
  Cob Construction 281

- Appendix AW
  3D Printed Buildings 284

Preface

The purpose of *Significant Changes to the International Residential Code®, 2021 Edition*, is to familiarize building officials, fire officials, plans examiners, inspectors, design professionals, contractors, and others in the building construction industry with many of the important changes in the 2021 *International Residential Code®* (IRC®). This publication is designed to assist code users in identifying the specific code changes that have occurred and understanding the reasons behind the changes. It is also a valuable resource for jurisdictions in their code-adoption process.

Only a portion of the code changes to the IRC are discussed in this book. The changes selected were identified for a number of reasons, including their frequency of application, special significance, or change in application. However, the importance of the changes not included is not to be diminished. Further information on all code changes can be found in the Complete Revision History to the 2021 I-Codes, available from the International Code Council® (ICC®) online store. This resource provides the published documentation for each successful code change contained in the 2021 IRC since the 2018 edition.

Significant Changes to the International Residential Code, 2021 Edition, is organized into nine parts, each representing a distinct grouping of code topics. It is arranged to follow the general layout of the IRC, including code sections and section number format. The table of contents, in addition to providing guidance in the use of this publication, allows for a quick identification of those significant code changes that occur in the 2021 IRC.

Throughout the book, each change is accompanied by a photograph, an application example or an illustration to assist and enhance the reader's understanding of the specific change. A summary and a discussion of the significance of the change are also provided. Each code change is identified by type, be it an addition, modification, clarification, or deletion.

The code change itself is presented in a legislative format similar to the style utilized for code-change proposals. Deleted code language is shown with a strikethrough, and new code text is indicated by underlining.

As a result, the actual 2021 code language is provided, as well as a comparison with the 2018 language, so the user can easily determine changes to the specific code text.

As with any code-change text, *Significant Changes to the International Residential Code, 2021 Edition*, is best used as a companion to the 2021 IRC. Because only a limited discussion of each change is provided, the code itself should always be referenced in order to gain a more comprehensive understanding of the code change and its application.

The commentary and opinions set forth in this text are those of the authors and do not necessarily represent the official position of ICC. In addition, they may not represent the views of any enforcing agency, as such agencies have the sole authority to render interpretations of the IRC. In many cases, the explanatory material is derived from the reasoning expressed by code-change proponents.

Comments concerning this publication are encouraged and may be directed to ICC at significantchanges@iccsafe.org.

About the *International Residential Code*®

Building officials, design professionals, contractors and others involved in the field of residential building construction recognize the need for a modern, up-to-date residential code addressing the design and installation of building systems through both prescriptive and performance requirements. The *International Residential Code*® (IRC), *2021 Edition*, is intended to meet these needs through model code regulations that safeguard the public health and safety in all communities, large and small. The IRC is kept up to date through ICC's open code-development process. The provisions of the 2018 edition, along with those code changes approved in the most recent code development cycle, make up the 2021 edition.

The IRC is one in a family of 15 International Codes® (I-Codes®) published by ICC. This comprehensive residential code establishes minimum regulations for residential building systems by means of prescriptive and performance-related provisions. It is founded on broad-based principles that make possible the use of new materials and new building designs. The IRC is a comprehensive code containing provisions for building, energy conservation, mechanical, fuel gas, plumbing and electrical systems. The IRC is available for adoption and use by jurisdictions internationally. Its use within a governmental jurisdiction is intended to be accomplished through adoption by reference, in accordance with proceedings established by the jurisdiction's laws.

Acknowledgments

Grateful appreciation is due to many ICC staff members, including those in Product Development, Publishing, Marketing, and Technical Services, for their generous assistance in the preparation of this publication. Fred Grable, P.E., ICC Senior Staff Engineer, shared his expertise and provided commentary on the plumbing provisions. Gregg Gress, formerly ICC Senior Technical Staff, provided welcome assistance on the mechanical and fuel gas provisions. John "Buddy" Showalter, P.E., ICC Senior Staff Engineer, reviewed the structural provisions giving helpful insight and suggestions. All contributed to the accuracy and quality of the finished product.

About the Authors

Stephen A. Van Note, CBO
International Code Council
Managing Director, Product Development

Stephen A. Van Note is the Managing Director of Product Development for the International Code Council (ICC), where he is responsible for developing technical resource materials in support of the International Codes. His role also includes the management, review, and technical editing of publications developed by ICC staff members and other expert authors. He has authored a number of ICC support publications, including *Residential Code Essentials and Inspector Skills.* In addition, Steve develops and presents *International Residential Code* seminars nationally. He has over 50 years of experience in the construction and building code arena. Prior to joining ICC in 2006, Steve was a building official for Linn County, Iowa. Prior to his 15 years at Linn County, he was a carpenter and construction project manager for residential, commercial, and industrial buildings. A certified building official and plans examiner, Steve also holds certifications in several inspection categories.

Sandra Hyde, P.E.
International Code Council
Senior Staff Engineer, Product Development

Sandra Hyde is a Senior Staff Engineer with the International Code Council (ICC), where, as part of the Product Development team, she develops technical resource materials in support of the structural provisions of the International Residential, Building and Existing Building Codes. Her role also includes review and technical editing of publications authored by ICC and engineering associations, and the presentation of technical seminars on the IRC and IBC structural provisions. She has authored and reviewed support publications, including *Significant Changes to the International Building Code, Special Inspection Manual,* and, in conjunction with APA, *Guide to the IRC Wall Bracing Provisions.* Prior to joining ICC in 2010, Sandra worked in manufacturing and research of engineered wood products. She is a Registered Civil Engineer in Idaho and California.

About the International Code Council®

The International Code Council is a nonprofit association that provides a wide range of building safety solutions including product evaluation, accreditation, certification, codification and training. It develops model codes and standards used worldwide to construct safe, sustainable, affordable and resilient structures. ICC Evaluation Service (ICC-ES) is the industry leader in performing technical evaluations for code compliance, fostering safe and sustainable design and construction.

Washington DC Headquarters:
500 New Jersey Avenue, NW, 6th Floor
Washington, DC 20001

Regional Offices:
Eastern Regional Office: (BIR)
Central Regional Office: (CH)
Western Regional Office: (LA)

Distribution Center (Lenexa, KS)

1-888-ICC-SAFE (1-888-422-7233)
www.iccsafe.org

Family of Solutions:

PART 1
Administration

Chapters 1 and 2

- **Chapter 1** Scope and Administration
- **Chapter 2** Definitions

R102.7.1
Additions, Alterations or Repairs

R202
Definition of Emergency Escape and Rescue Opening

R202
Definition of Townhouse

The administration part of the *International Residential Code* (IRC) covers the general scope, purpose, applicability, and other administrative issues related to the regulation of residential buildings by building safety departments. The administrative provisions establish the responsibilities and duties of the various parties involved in residential construction and the applicability of the technical provisions within a legal, regulatory, and code-enforcement arena.

Section R101.2 establishes the criteria for buildings that are regulated by the IRC. Buildings beyond the scope of Section R101.2 are regulated by the *International Building Code* (IBC). The remaining topics in the administration provisions of Chapter 1 include the establishment of the building safety department, duties of the building official, permits, construction documents, and inspections.

The definitions contained within the IRC are intended to reflect the special meaning of such terms within the scope of the code. As terms can often have multiple meanings within their ordinary day-to-day use or within the various disciplines of the construction industry, it is important that their meanings within the context of the IRC be understood. Most definitions used throughout the IRC are found in Chapter 2, but additional definitions specific to the applicable topics are found in the energy provisions of Chapter 11, the fuel gas provisions of Chapter 24, and the electrical provisions of Chapter 35. ■

R102.7.1
Additions, Alterations or Repairs

CHANGE TYPE: Clarification

CHANGE SUMMARY: The code references the *International Existing Building Code®* (IEBC®) only when alterations are part of a change to a use or occupancy outside the scope of the IRC.

2021 CODE: R102.7.1 Additions, alterations or repairs. Additions, alterations or repairs to any structure shall conform to the requirements for a new structure without requiring the existing structure to comply with the requirements of this code, unless otherwise stated. Additions, alterations, repairs and relocations shall not cause an existing structure to become ~~unsafe or adversely affect the performance of the building.~~ <u>less compliant with the provisions of this code than the existing building or structure was prior to the addition, alteration or repair. An existing building together with its additions shall comply with the height limits of this code. Where the alteration causes the use or occupancy to be changed to one not within the scope of this code, the provisions of the *International Existing Building Code* shall apply.</u>

CHANGE SIGNIFICANCE: As with other international codes, the IRC provides relief for existing buildings to allow the legal occupancy to continue without fully complying with current codes. To impose regulations to bring existing buildings into current compliance would be impractical and unreasonable. This provision also applies to existing buildings undergoing modifications or additions. Generally, only the modification or addition need comply with the current code. Previous editions of the IRC stated that additions, alterations or repairs could not cause any portion of the existing building to become unsafe or otherwise adversely affect the performance of the building. For clarification, that provision has changed to state that modifications cannot cause the existing building to become less compliant with the

Addition to a single-family dwelling.

current code. Additional language emphasizes that the building height limitations still apply to additions. The relationship of the IEBC to the IRC has caused some confusion and is clarified in the added language. The IEBC offers alternative compliance paths for renovations to existing buildings. The code now only sends users to the IEBC if the alteration or addition is part of a change of use or change of occupancy that takes the building outside the scope of the IRC. That is, the new use is classified as a building under the IBC and the IEBC applies to the modifications. Similarly, the IEBC states in an exception to its scope that modifications to one- and two-family dwellings and townhouses and their accessory buildings are regulated either by the IEBC or the IRC. Appendix J of the IRC also offers guidance and alternatives for compliance with the code during renovation of existing buildings. The appendix chapters are only in effect if specifically adopted by the jurisdiction. Appendix J provisions, similar to those found in the IEBC, intend to encourage the continued use or reuse of legally existing buildings and structures.

R202

Definition of Emergency Escape and Rescue Opening

CHANGE TYPE: Clarification

CHANGE SUMMARY: Definitions for emergency escape and rescue openings and grade floor openings have been updated for clarification and to be consistent with the IBC.

2021 CODE: R202 DEFINITIONS

EMERGENCY ESCAPE AND RESCUE OPENING. An operable exterior window, door or <u>other</u> similar device that provides for a means of escape and access for rescue in the event of an emergency. (See also "Grade floor <u>emergency escape and rescue</u> opening.")

GRADE FLOOR <u>EMERGENCY ESCAPE AND RESCUE</u> OPENING. ~~A window or other~~ <u>An emergency escape and rescue</u> opening located such that the ~~sill height~~ <u>bottom</u> of the <u>clear</u> opening is not more than 44 inches (1118 mm) above or below the finished ground level adjacent to the opening. (See also "Emergency escape and rescue opening.")

CHANGE SIGNIFICANCE: The emergency escape and rescue provisions have regularly undergone revisions in past code cycles and the same is true of the 2021 IRC. In conjunction with the reorganization and new text in Section R310, two related definitions have been revised. In part, these changes are for consistency with the IBC and other I-Codes. They also clarify their meaning and application. In the 2012 IRC, the measurement for the maximum height of the emergency escape opening (often referred to as the sill height) was clearly spelled out as the distance from the finished floor to the lowest point of the net clear opening. Traditionally that

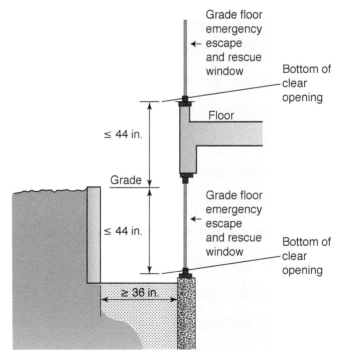

Grade floor emergency escape and rescue openings require only 5.0 sq. ft. of clear area.

has been interpreted and used as the intent of the code, even when the terminology was more ambiguous. The measurement language of the 2012 code was dropped in the 2015 and 2018 editions. Now the definition clearly spells out that the measurement is taken to the lowest point of the opening, which may or may not be a traditional windowsill.

There have been similar misunderstandings related to the definition of grade floor opening. Section R310 has always allowed a reduction in the area of the net clear opening for emergency escape and rescue openings close to grade. This is based on the reduced hazard when compared to escape and rescue from a second or third story opening. However, the definition previously did not mention emergency escape and rescue and its connection to Section R310 was not apparent. As a result, there has been some confusion as to when the exception for reduced opening area was in effect. The code stated that the exception applied to grade floor openings and below grade openings. A literal interpretation of this language would allow a below grade opening deeper than 44 inches below grade and still take advantage of the size reduction, though that was not the intent. To make the connection clear, the definition has been revised to include emergency escape and rescue in the term.

R202
Definition of Townhouse

CHANGE TYPE: Clarification

CHANGE SUMMARY: A revised definition of townhouse (a building) and a new definition for townhouse unit (a dwelling unit) clarify the appropriate use of the terms.

2021 CODE: R202 DEFINITIONS

BUILDING. Any one- or two-family dwelling <u>or townhouse</u>, or portion thereof, ~~including townhouses,~~ used or intended to be used for human habitation, for living, sleeping, cooking or eating purposes, or any combination thereof, or any accessory structure. For the definition applicable in Chapter 11, see Section N1101.6.

TOWNHOUSE. A ~~single-family dwelling unit constructed in a group of~~ <u>building that contains</u> three or more attached <u>townhouse</u> units. ~~in which each unit extends from foundation to roof and with a yard or public way on not less than two sides.~~

TOWNHOUSE UNIT. <u>A single-family dwelling unit in a townhouse that extends from foundation to roof and that has a yard or public way on not less than two sides.</u>

CHANGE SIGNIFICANCE: There has been some confusion and certainly some inconsistency in the use of the term "townhouse." Previously, a townhouse was defined as a single-family dwelling unit in a group of three or more dwelling units in one building. However, the term was used interchangeably to describe the entire building and the individual dwelling units within that building. The new definition of "townhouse unit" intends to remedy that inconsistency. A townhouse unit describes each individual single-family dwelling unit in a townhouse building. Therefore, a townhouse is a building that contains three or more townhouse units. The new terms appear in the townhouse provisions of Section R302.2 and in other locations throughout the code. The definition of building has also been updated to make a direct reference to townhouse in addition to one- and two-family dwellings and accessory structures.

Townhouse.

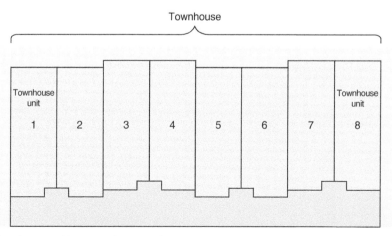

Townhouse with 8 townhouse units.

PART 2
Building Planning
Chapter 3

- Chapter 3   Building Planning

Chapter 3 includes the bulk of the nonstructural provisions, including the location on the lot, fire-resistant construction, light and ventilation, emergency escape and rescue, fire protection, safety glazing, fall protection, and many other provisions aimed at protecting the health, safety, and welfare of the public. In addition to such health and life-safety issues, Chapter 3 provides the overall structural design criteria for residential buildings regulated by the IRC. Section R301 addresses live loads, dead loads, and environmental loads such as wind, seismic, and snow.

R301.1.4
Intermodal Shipping Containers

R301.2
Wind Speeds

TABLE R301.2.1(1)
Component and Cladding Wind Pressures

R301.2.1.1
Special Wind Regions

R301.2.2.6
Irregular Buildings in Seismic Areas

R301.2.2.10
Seismic Anchorage of Water Heaters

R301.3
Story Height

R302.2
Townhouses

R302.3
Two-Family Dwelling Separation

R302.4
Dwelling Unit Rated Penetrations

R302.5
Dwelling-Garage Opening Protection

R303.1
Mechanical Ventilation

R305.1
Ceiling Height

R308.4.5
Glazing and Wet Surfaces

R308.6
Skylight Glass Retention Screens

R310.1
Emergency Escape and Rescue Opening Required

R310.2
Emergency Escape and Rescue Openings

R310.3, R310.4
Area Wells for Emergency Escape and Rescue Openings

R310.5, R310.6, R310.7
Emergency Escape and Rescue Openings in Existing Buildings

R311.7, R311.8
Stairways and Ramps

R311.7.7
Stairway and Landing Walking Surface

R312.2
Window Fall Protection

R314.3
Smoke Alarm Locations

R315.2.2
Carbon Monoxide Alarms

R317.1
Protection of Wood Against Decay

R320
Accessibility

R323
Storm Shelters

R324.3
Photovoltaic Systems

R326
Habitable Attics

R301.1.4 Intermodal Shipping Containers

CHANGE TYPE: Addition

CHANGE SUMMARY: Provisions for construction with intermodal shipping containers are added to the *International Residential Code* (IRC).

2021 CODE TEXT: <u>**R301.1.4 Intermodal shipping containers.** Intermodal shipping containers that are repurposed for use as buildings or structures, shall be designed in accordance with the structural provisions in Section 3115 of the *International Building Code.*</u>

CHANGE SIGNIFICANCE: A wide variety of materials are regulated throughout the IRC. In addition to typical homes constructed of wood, other types of construction are addressed by reference to other codes or standards. Section R104.11 allows for the use of alternative materials and methods of construction provided such methods and materials have been approved by the building official. The use of intermodal shipping containers as buildings and structures is now specifically recognized in the IRC and criteria have been established to address the minimum safety requirements by reference to Section 3115 of the *International Building Code* (IBC). Additionally, ICC G5-2019 *Guideline for the Safe Use of ISO Shipping Containers Repurposed as Buildings and Building Components* was recently published to assist building departments in their evaluation.

Over 30 million intermodal shipping containers are in use around the world today. These containers, both new and used, are being repurposed and converted to occupiable structures. About 80 percent of shipping containers are either 20-foot (6.1 m) or 40-foot (12.2 m) standard length boxes for dry freight. These typical containers are rectangular, closed boxes, with doors fitted at one end, and made of corrugated weathering steel with a plywood floor. The corrugating of the sheet metal used for the sides and roof contributes significantly to the container's rigidity and

Single container home.

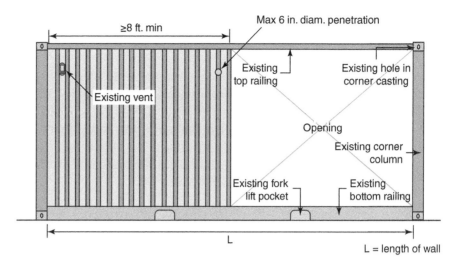

Maximum penetration size in shipping containers used for bracing.

stacking strength. Standard containers are 8-foot (2.44 m) wide by 8.5 feet (2.59 m) tall, although the taller hi-cube units measuring 9.5 feet (2.90 m) tall have become common in recent years.

The containers may sit at yards waiting to be used or can be appropriated for use as building materials. Like any repurposed material, they must be evaluated for strength and condition. By referencing Section 3115 of the IBC, the IRC brings in requirements for inspection of the containers before use to ensure material quality. The IBC requires an engineered design for the use of the containers. This design assumes that the containers meet the requirements of ISO 1496-1 which details testing for the strength capacity of the containers. All containers will be tested by an approved third-party to verify that they still meet the capacity of the ISO standard. This testing includes a check of the strength of the side walls, end walls, floor and roof; the rigidity of the container; and its ability to be lifted and stacked. When containers will be used individually in Seismic Design Categories (SDC) A, B or C, they may meet a simplified list of structural criteria.

R301.2 Wind Speeds

CHANGE TYPE: Modification

CHANGE SUMMARY: Updated wind speed maps match IBC and ASCE 7 maps with a large portion of the country having wind speeds less than 115 mph.

2021 CODE TEXT: R301.2.1 Wind design criteria. Buildings and portions thereof shall be constructed in accordance with the wind provisions of this code using the ultimate design wind speed in Table ~~R301.2(1)~~ R301.2 as determined from Figure ~~R301.2(5)A~~ R301.2(2). The structural provisions of this code for wind loads are not permitted where wind design is required as specified in Section R301.2.1.1. Where different construction methods and structural materials are used for various portions of a building, the applicable requirements of this section for each portion shall apply. Where not otherwise specified, the wind loads listed in Table ~~R301.2(2)~~ R301.2.1(1) adjusted for height and exposure using Table ~~R301.2(3)~~ R301.2.1(2) shall be used to determine design

Wind damage.

Location	Vmph	(m/s)
Guam	195	(87)
Virgin Islands	165	(74)
American Samoa	160	(72)
Hawaii – Special Wind Region Statewide	130	(58)

Notes:
1. Values are nominal design 3-second gust wind speeds in miles per hour (m/s) at 33 ft (10 m) above ground for Exposure C category.
2. Linear interpolation is permitted between contours. Point values are provided to aid with interpolation.
3. Islands, coastal areas, and land boundaries outside the last contour shall use the last wind speed contour.
4. Mountainous terrain, gorges, ocean promontories, and special wind regions shall be examined for unusual wind conditions.
5. Wind speeds correspond to approximately a 7% probability of exceedance in 50 years (Annual Exceedance Probability = 0.00143, MRI = 700 Years).
6. Location-specific basic wind speeds shall be permitted to be determined using www.atcouncil.org/windspeed.

Updated Wind Speeds.

load performance requirements for wall coverings, curtain walls, roof coverings, exterior windows, skylights, garage doors and exterior doors. Asphalt shingles shall be designed for wind speeds in accordance with Section R905.2.4. <u>Metal roof shingles shall be designed for wind speeds in accordance with Section R905.4.4.</u> A continuous load path shall be provided to transmit the applicable uplift forces in Section ~~R802.11.1~~ <u>R802.11</u> from the roof assembly to the foundation. <u>Where ultimate design wind speeds in Figure R301.2(2) are less than the lowest wind speed indicated in the prescriptive provisions of this code, the lowest wind speed indicated in the prescriptive provision of this code shall be used.</u>

CHANGE SIGNIFICANCE: Section R301.2.1 coordinates the IRC wind design criteria with the 2016 edition of the engineering standard *Minimum Design Loads and Associated Criteria for Buildings and Other Structures* (ASCE 7). In ASCE 7-16, wind speeds in non-hurricane prone areas of the contiguous United States have been revised using contours to better reflect regional variations in extreme straight-line winds due to thunderstorms.

In Figure R301.2(2), wind speeds are no longer a minimum of 115 mph for the center of the country and 110 mph in the west. The map is updated to show lower wind speeds with isolines for 90, 95, 100 and 105 mph. Point values are added to the map to aid interpolation between isolines. Generally, wind speeds have dropped across the country, and in some locations the wind speed dropped significantly. Any area that had wind speeds set at 110 mph (west coast) or 115 mph (central United States) now has reduced wind speeds.

With updates to Figure R301.2(2), the map is now identical to the 2021 IBC and ASCE 7-16 wind speed maps for Risk Category II buildings – the category for most buildings including single- and two-family residences and townhouses. Wind speeds in hurricane-prone regions generally remained the same. For the northeastern United States, certain wind speeds dropped 5 to 10 mph inland away from the coastline. New hurricane contours were developed based on updated hurricane models, and hurricane coastline contour locations were adjusted to reflect new research into hurricane decay rates over land. The details of changes, data behind the isolines and methods used to estimate both non-hurricane and hurricane wind speeds are provided in ASCE 7-16's Commentary to Chapter 26. Note that while wind speeds have decreased in certain parts of the country, component and cladding roof wind pressures in certain cases have increased due to changes in Table R301.2.1(1). See the significant change discussion for roof components and cladding.

To see a specific wind speed for a town or individual building, go to either hazards.atcouncil.org or asce7hazardtool.online and type in an address or GPS coordinates. The website will give the wind speed assigned to the location. It is now possible to determine the ground snow load, wind speed, seismic design category and tornado risk from the Applied Technology Council (ATC) website, which remains free to users. The American Society of Civil Engineers (ASCE) website contains additional information while charging a nominal yearly fee and offering wind speeds and tsunami hazard zones for free.

Section R301.2.1 now also includes a reference for wind design of metal roof shingles. Metal roof shingles are fastened following the requirements of Section R905.4.4.

Table R301.2.1(1) — Component and Cladding Wind Pressures

CHANGE TYPE: Modification

CHANGE SUMMARY: Component and cladding wind pressures in Table R301.2.1(1) are updated for new design wind speeds and hip or gable roof profiles.

2021 CODE TEXT:

TABLE R301.2.1(1): Component and Cladding Loads for a Building with a Mean Roof Height of 30 Feet Located in Exposure B (ASD) (psf)

	Zone	Effective Wind Areas (ft²)	Ultimate Design Wind Speed, V_{ult}											
			90		95		100		105		110		...	180	
			Pos	Neg	Pos	Neg	Pos	Neg	Pos	Neg	Pos	Neg	Pos Neg	Pos	Neg
Gable Roof >7 to 20 degrees	1, 2e	10	5.4	-16.2	6	-18.0	6.7	-19.9	7.4	-22	8.1	-24.1	... ...	21.6	-64.6
	1, 2e	20	4.9	-16.2	5.4	-18	6.0	-19.9	6.6	-22	7.2	-24.1	... ...	19.4	-64.6
	1, 2e	50	4.1	-9.9	4.6	-11	5.1	-12.2	5.6	-13.4	6.1	-14.7	... ...	16.4	-39.4
	1, 2e	100	3.6	-5	4	-5.6	4.4	-6.2	4.8	-6.9	5.3	-7.5	... ...	14.2	-20.2
	2n, 2r, 3e	10	5.4	-23.6	6	-26.3	6.7	-29.1	7.4	-32.1	8.1	-35.2	... ...	21.6	-94.2
	2n, 2r, 3e	20	4.9	-20.3	5.4	-22.7	6	-25.1	6.6	-27.7	7.2	-30.4	... ...	19.4	-81.4
	2n, 2r, 3e	50	4.1	-16	4.6	-17.9	5.1	-19.8	5.6	-21.8	6.1	-24	... ...	16.4	-64.2
	2n, 2r, 3e	100	3.6	-12.8	4	-14.3	4.4	-15.8	4.8	-17.4	5.3	-19.1	... ...	14.2	-51.3
	3r	10	5.4	-28	6	-30.2	6.7	-34.6	7.4	-38.1	8.1	-41.8	... ...	21.6	-112
	3r	20	4.9	-24	5.4	-26.7	6	-29.6	6.6	-32.7	7.2	-35.9	... ...	19.4	-96
	3r	50	4.1	-18.7	4.6	-20.8	5.1	-23.1	5.6	-25.4	6.1	-27.9	... ...	16.4	-74.7
	3r	100	3.6	-14.7	4	-16.3	4.4	-18.1	4.8	-20	5.3	-21.9	... ...	14.2	-58.7
Hipped Roof >7 to 20 degrees	1	10	6.5	-14.7	7.3	-16.3	8	-18.1	8.9	-20	9.7	-21.9	... ...	26.1	-58.7
	1	20	5.6	-14.7	6.3	-16.3	7	-18.1	7.7	-20	8.4	-21.9	... ...	22.5	-58.7
	1	50	4.4	-11.3	5	-12.6	5.5	-14	6.1	-15.4	6.6	-16.9	... ...	17.8	-45.3
	1	100	3.6	-8.7	4	-9.7	4.4	-10.8	4.8	-11.9	5.3	-13.1	... ...	14.2	-35
	2r	10	6.5	-19.1	7.3	-21.3	8	-23.6	8.9	-26	9.7	-28.6	... ...	26.1	-76.5
	2r	20	5.6	-17.2	6.3	-19.2	7	-21.3	7.7	-23.4	8.4	-25.7	... ...	22.5	-68.9
	2r	50	4.4	-14.7	5	-16.4	5.5	-18.2	6.1	-20	6.6	-22	... ...	17.8	-58.8
	2r	100	3.6	-12.8	4	-14.3	4.4	-15.8	4.8	-17.4	5.3	-19.1	... ...	14.2	-51.3
	2e, 3	10	6.5	-20.6	7.3	-22.9	8	-25.4	8.9	-28	9.7	-30.8	... ...	26.1	-82.4
	2e, 3	20	5.6	-18.5	6.3	-20.6	7	-22.9	7.7	-25.2	8.4	-27.7	... ...	22.5	-74.1
	2e, 3	50	4.4	-15.8	5	-17.6	5.5	-19.5	6.1	-21.5	6.6	-23.6	... ...	17.8	-63.1
	2e, 3	100	3.6	-13.7	4	-15.3	4	-16.9	4.8	-18.7	5.3	-20.5	... ...	14.2	-54.8

(Only a portion of the table is shown for brevity and clarity.)

CHANGE SIGNIFICANCE: Changes to Section R301.2 coordinate wind design criteria in the IRC with the referenced engineering load standard *Minimum Design Loads and Associated Criteria for Buildings and Other Structures* (ASCE 7-16). Simplified component and cladding loads in Table R301.2.1(1) are revised for consistency with ASCE 7 roof component and cladding loads (C&C) for buildings with mean roof heights less than or equal to 60 feet. The roof zones and pressure coefficients in ASCE 7-16 Figure 30.3-2 (which includes Figures 30.3-2A through 30.3-2I) have been revised based on analysis of an extensive wind tunnel test results database.

Compared to previous versions of the IRC, C&C pressure coefficients have increased. C&C roof zone sizes are also modified. Monitoring of buildings across the country indicates that for low-rise buildings, C&C roof zone sizes depend primarily on building height, h. Note that for Exposure B, when the building mean roof height is less than 30 feet, the adjustment is less than 1.0 allowing a reduction in required wind pressure. Mean roof height is defined as the average of the ridge and eave heights.

Roof cladding damage.

Photo courtesy of MICHELANGELOBOY/E+/Getty Images

TABLE R301.2.1(2) Height and Exposure Adjustment Coefficients for Table R301.2.1(1)

Mean Roof Height (ft)	Exposure		
	B	C	D
15	0.82 ~~1.00~~	1.21	1.47
20	0.89 ~~1.00~~	1.29	1.55
25	0.94 ~~1.00~~	1.35	1.61
30	1.00	1.40	1.66
35	1.05	1.45	1.70

Figure R301.2.1, component and cladding pressure zones, is illustrated in the figure on gable roof wind zones and shows corner (3, 3e, 3r), edge (2e, 2r, 2n) and interior (1) roof zones. These C&C zones are different from roof zones in previous editions of the IRC. The updated Figure R301.2.1 and Table R301.2.1(1) incorporate recent research by increasing edge and corner wind pressures as appropriate. To better define which roof surface areas require increased wind resistance, gable and hip roofs are divided into two categories and low-slope roofs (0 to 7 degrees) are separated from roofs with shallow slopes (>7 to 20 degrees) and steeper slopes (>20 to 27) and (>27 to 45 degrees). By separating the roof slope into multiple categories and dividing the roof surface into multiple regions, nailing patterns are increased only when necessary and less restrictive patterns can be used where appropriate.

New vocabulary includes division of C&C corner and edge zones as follows:
2 – edge zones
    2e – edge zone along bottom of roof above the soffit
    2r – edge zone along roof peak
    2n – edge zone along rake edge of gable roof

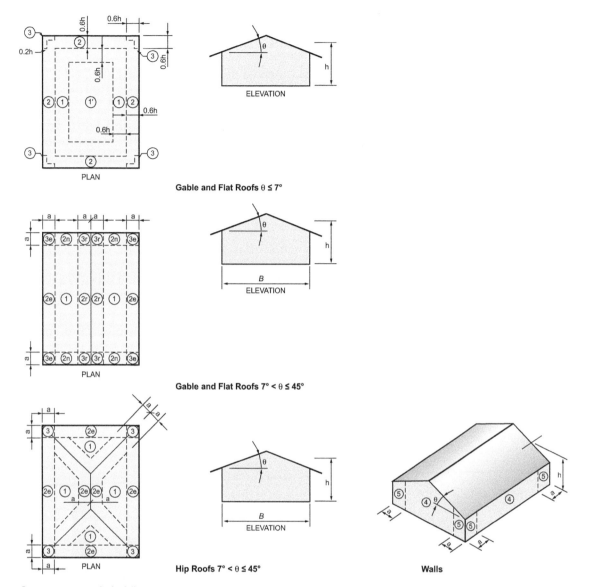

Component and cladding pressure zones.

3 – corner zones
    3e – corner zone at bottom of roof above the soffit
    3r – corner zone at roof peak
    a = 4 feet

Interior C&C zones are broken into two categories. For the IRC, zone 1 and zone 1' use the same value. If the roof requires design per the IBC, these values will be different.
    1 – interior zone
    1' – central interior zone, flat or low slope roof

When considering nailing patterns for buildings constructed following the 2021 IRC, consider how many different nailing patterns are reasonable to require on a single roof. A single nailing pattern is preferred by framers, but in high wind zones, it may be preferable to have a nailing pattern for corner and edge zones with a different pattern in the interior of the roof. Also note that relatively new fasteners, such as Roof Sheathing Ring Shank (RSRS) nails have been tabulated in Table R602.3(1) specifically to address these increased roof wind pressures.

R301.2.1.1
Special Wind Regions

CHANGE TYPE: Clarification

CHANGE SUMMARY: Engineered design requirements for special wind regions are explicitly stated in Section R301.2.1.1.

2021 CODE TEXT: **R301.2.1.1 Wind limitations and wind design required.** The wind provisions of this code shall not apply to the design of buildings where wind design is required in accordance with ~~Figure R301.2(5)B~~ <u>Figure R301.2.1.1, or where the ultimate design wind speed, V_{ult}, in Figure R301.2(2) equals or exceeds 140 mph in a special wind region</u>.

Exceptions:

1. For concrete construction, the wind provisions of this code shall apply in accordance with the limitations of Sections R404 and R608.
2. For structural insulated panels, the wind provisions of this code shall apply in accordance with the limitations of Section R610.
3. For cold-formed steel light-frame construction, the wind provisions of this code shall apply in accordance with the limitations of Sections R505, R603 and R804.

High winds in a special wind region.

In regions where wind design is required in accordance with ~~Figure R301.2(5)B~~ Figure R301.2.1.1 <u>or where the ultimate design wind speed, V_{ult}, in Figure R301.2(2) equals or exceeds 140 mph in a special wind region</u>, the design of buildings for wind loads shall be in accordance with one or more of the following methods:

1. AWC *Wood Frame Construction Manual* (WFCM).
2. ICC *Standard for Residential Construction in High-Wind Regions* (ICC 600).
3. ASCE *Minimum Design Loads for Buildings and Other Structures* (ASCE 7).
4. AISI *Standard for Cold-Formed Steel Framing—Prescriptive Method for One- and Two- Family Dwellings* (AISI S230).
5. *International Building Code.*

The elements of design not addressed by the methods in Items 1 through 5 shall be in accordance with the provisions of this code.
Where ASCE 7 or the *International Building Code* is used for the design of the building, the wind speed map and exposure category requirements as specified in ASCE 7 and the *International Building Code* shall be used.

CHANGE SIGNIFICANCE: The *2018 International Residential Code* (IRC) did not explicitly prohibit IRC prescriptive provisions in special wind regions for wind speeds less than 140 mph. Rather, special wind regions were identified in the map for areas with wind design required in hurricane prone regions but no text stated whether the special wind region also required wind design. The 2021 IRC clarifies the intent of the code by stating that the IRC prescriptive provisions can be used in special wind regions where the wind speed is less than 140 mph. If design wind speeds are 140 mph or greater, an alternative design method must be followed.

Figure R301.2.1.1 prohibits use of IRC prescriptive provisions for the structural frame of the building in shaded areas of the map which are defined as "wind design required" regions. These regions mainly cover hurricane-prone areas where wind speeds exceed 130 mph (gulf coast and southern Atlantic coast, Caribbean, South Pacific and Hawaii) and other coastal areas where wind speeds exceed 140 mph (northern and central Atlantic coast and Alaska). Special wind regions are also defined on the map, but a special wind region has highly variable wind speeds depending on specific lot locations. Not all wind speeds in these regions are above 140 mph. With this clarification, it is clear that engineering is only required in special wind regions where design wind speeds exceed 140 mph.

Engineered design is also required for buildings on upper slopes of hills when design wind speeds exceed 140 mph. Table R301.2.1.5.1, Ultimate Design Wind Speed Modification for Topographic Wind Effect, now lists wind speeds for buildings on the upper half of a hill for lower design wind speed areas. Values have been added for design wind speeds of 95, 100 and 105 mph.

R301.2.2.6
Irregular Buildings in Seismic Areas

CHANGE TYPE: Addition

CHANGE SUMMARY: Irregular building limitations now include hillside light-frame construction.

2021 CODE TEXT: R301.2.2.6 Irregular buildings. The seismic provisions of this code shall not be used for structures, or portions thereof, located in Seismic Design Categories C, D_0, D_1 and D_2 and considered to be irregular in accordance with this section. A building or portion of a building shall be considered to be irregular where one or more of the conditions defined in Items 1 through ~~7~~ 8 occur. Irregular structures, or irregular portions of structures, shall be designed in accordance with accepted engineering practice to the extent the irregular features affect the performance of the remaining structural system. Where the forces associated with the irregularity are resisted by a structural system designed in accordance with accepted engineering practice, the remainder of the building shall be permitted to be designed using the provisions of this code.

(No changes to Items 1-7)

> 8. <u>**Hillside Light-Frame Construction.** Conditions in which all of the following apply:</u>
>   8.1. <u>The grade slope exceeds 1 vertical in 5 horizontal where averaged across the full length of any side of the dwelling.</u>

Irregular building shape.

8.2. The tallest cripple wall clear height exceeds 7 feet (2134 mm), or where a post and beam system occurs at the dwelling perimeter, the post and beam system tallest post clear height exceeds 7 feet (2134 mm),

8.3. Of the total plan area below the lowest framed floor, whether open or enclosed, less than 50 percent is living space having interior wall finishes conforming to Section R702.

Where Item 8 is applicable, design in accordance with accepted engineering practice shall be provided for the floor diaphragm immediately above the cripple walls or post and beam system and all structural elements and connections from this diaphragm down to and including connections to the foundation and design of the foundation to transfer lateral loads from the framing above.

Exception: Light-frame construction in which the lowest framed floor is supported directly on concrete or masonry walls over the full length of all sides except the downhill side of the dwelling need not be considered an irregular dwelling under Item 8.

R202 CRIPPLE WALL CLEAR HEIGHT. The vertical height of a cripple wall from the top of the foundation to the underside of floor framing above.

CHANGE SIGNIFICANCE: For light-frame dwellings on steep hillsides, the typical assumption of floor loads transferring to braced wall panels based on the tributary area of a flexible wood floor does not work for adequate seismic performance. Whether earthquake shaking is across the slope or perpendicular to the hill, seismic forces follow the stiffest load path to the uphill foundation, rather than distributing evenly to all braced wall panels as assumed in IRC seismic wall bracing provisions. To address this issue, a trigger was added to Section R301.2.2.6 requiring engineered design of hillside dwellings, now defined as irregular buildings.

Hillside dwellings were found to be vulnerable in the 1994 Northridge, California earthquake. 117 significantly damaged hillside dwellings of typical light-frame wood construction were identified in reconnaissance reporting, with an additional 40 damaged buildings utilizing post and beam foundations. 15 dwellings were reported to have collapsed with another 15 near collapse.

The slope trigger (Item 8.1) is used to limit applicability of this irregularity to dwellings that are on sites with a significant slope. Averaging the grade along the side of the dwelling is intended to focus on the overall elevation drop across the dwelling and not trigger the irregularity based only on limited areas of steeper slope. For most dwellings this criterion will be evaluated by looking at the grade elevation on each side of the building. For large and complex dwellings, additional "sides" will need to be evaluated.

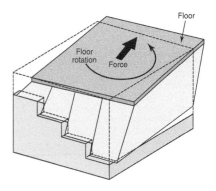

Seismic forces across the slope cause the building to rotate.

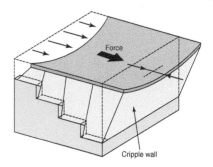

Seismic forces in the direction of the slope move the building downhill causing the center of the foundation to bend or flex inward.

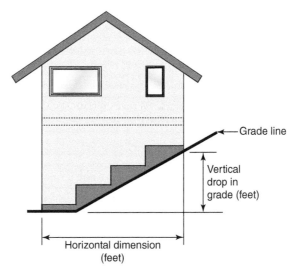

Determine slope by dividing the elevation change by the slope.

A second trigger considers cripple wall height. A cripple wall height or post height greater than seven feet tall triggers an engineered design (Item 8.2). Lastly, where a basement area is less than 50 percent finished and not sheathed with interior finish (Item 8.3), engineering of the bracing system is required. Interior finish stiffens walls, increasing a building's ability to resist earthquake forces. All three triggers must be met before a dwelling is deemed irregular. These triggers were observed as points at which damage and displacements of uphill foundations appear to significantly increase the likelihood of building collapse.

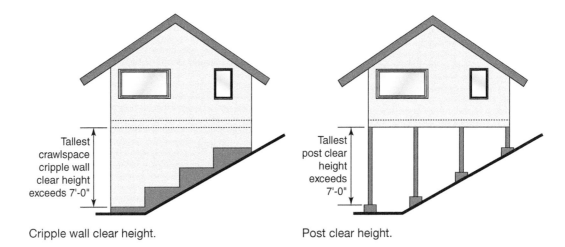

Cripple wall clear height.   Post clear height.

The exception exempts dwellings that have full-height concrete or masonry walls. For a dwelling with a simple rectangular floor plan, full-height concrete or masonry walls would need to occur on three sides (excluding the downhill side) to qualify for the exception. For a more complex dwelling configuration, additional concrete or masonry walls would be required to qualify. Dwellings with doors and windows in the concrete or masonry walls still qualify for the exception. In all dwellings, the concrete or masonry walls need to conform to applicable IRC provisions. A basement described in the exception could have a wood cripple wall as the downhill side exterior building wall.

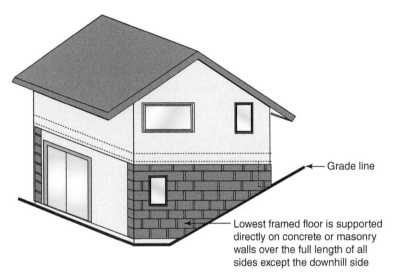

Basement with masonry walls on three sides.

R301.2.2.10
Seismic Anchorage of Water Heaters

CHANGE TYPE: Modification

CHANGE SUMMARY: Water heaters and thermal storage units in Townhouses in SDC C must be anchored.

2021 CODE TEXT: R301.2.2.10 Anchorage of water heaters. In Seismic Design Categories D_0, D_1 and D_2, <u>and in townhouses in Seismic Design Category C,</u> water heaters <u>and thermal storage units</u> shall be anchored against movement and overturning in accordance with Section ~~M1307~~ <u>M1307.2 or P2801.8</u>.

CHANGE SIGNIFICANCE: Section M1307.2, Anchorage of appliances, defines appliances as both water heaters and thermal storage tanks. The section applies not only to Seismic Design Categories (SDC) D_0, D_1 and D_2 but also to townhouses in SDC C. For consistency, Section R301.2.2.10 now requires both water heaters and thermal storage units to be anchored in SDC C (for townhouses only) and SDC D_0, D_1 and D_2 while a reference is added to the appropriate plumbing section.

Anchorage is required because a water heater can move around a floor or off a platform during an earthquake. Due to their weight, water heaters can cause damage by falling and be a flood hazard. If gas-powered, movement of a water heater may break a gas line, leading to a fire. Thermal storage units, similar to electric water heaters, consist of a tank of water heated by solar collectors. A thermal storage tank may be located on a roof, in a garage, attic or crawl space. This tank presents the same fall and flood hazard as an electric water heater and must also be anchored against movement during an earthquake.

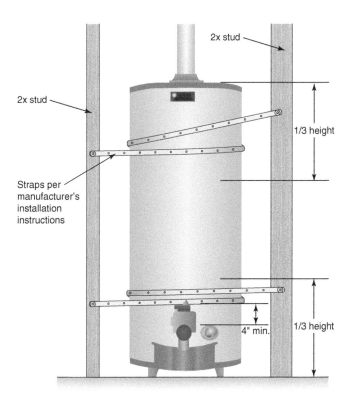

Anchorage of water heater.

Significant Changes to the IRC 2021 Edition   R301.3 ■ Story Height

R301.3
Story Height

CHANGE TYPE: Clarification

CHANGE SUMMARY: Maximum story height for wood wall framing is 13 feet 7 inches when the exception requirements are met.

2021 CODE TEXT:

R301.3 Story height. The wind and seismic provisions of this code shall apply to buildings with story heights not exceeding the following:

1. For wood wall framing, the story height shall not exceed 11 feet 7 inches (3531 mm) and the laterally unsupported bearing wall stud height permitted by Table R602.3(5).

   Exception: <u>A story height not exceeding 13 feet 7 inches (4140 mm) is permitted provided the maximum wall stud clear height does not exceed 12 feet (3658 mm), the wall studs are in accordance with</u>

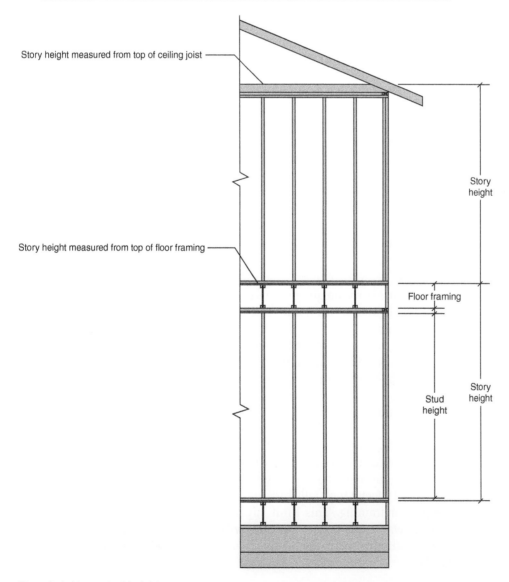

Story height vs. stud height.

Exception 2 or Exception 3 of Section R602.3.1 or an engineered design is provided for the wall framing members, and wall bracing for the building is in accordance with Section R602.10. Studs shall be laterally supported at the top and bottom plate in accordance with Section R602.3.

2. For cold-formed steel wall framing, the story height shall be not more than 11 feet 7 inches (3531 mm) and the unsupported bearing wall stud height shall be not more than 10 feet (3048 mm).

3. For masonry walls, the story height shall be not more than 13 feet 7 inches (4140 mm) and the bearing wall clear height shall be not more than 12 feet (3658 mm).

Exception: An additional 8 feet (2438 mm) of bearing wall clear height is permitted for gable end walls.

4. For insulating concrete form walls, the maximum story height shall not exceed 11 feet 7 inches (3531 mm) and the maximum unsupported wall height per story as permitted by Section R608 tables shall not exceed 10 feet (3048 mm).

5. For structural insulated panel (SIP) walls, the story height shall be not more than 11 feet 7 inches (3531 mm) and the bearing wall height per story as permitted by Section R610 tables shall not exceed 10 feet (3048 mm).

For walls other than wood-framed walls, individual walls or wall studs shall be permitted to exceed these limits as permitted by Chapter 6 provisions, provided that the story heights of this section are not exceeded. An engineered design shall be provided for the wall or wall framing members where the limits of Chapter 6 are exceeded. Where the story height limits of this section are exceeded, the design of the building, or the noncompliant portions thereof, to resist wind and seismic loads shall be in accordance with the *International Building Code.*

CHANGE SIGNIFICANCE: Section R301.3 is updated to address confusion with story height provisions. In the 2003 and 2006 IRC, Section R301.3 allowed wood-frame buildings to have maximum bearing wall stud heights of 10 feet supporting floor framing not exceeding 16 inches in depth. An exception allowed a maximum bearing wall stud height of 12 feet provided an engineered design for the wall was provided for everything other than wind and seismic wall bracing, which could be determined per Section R602.10 with adjustment factors to increase the bracing amounts for taller walls.

In the 2009 IRC, revision of Section R301.3 allowed floor framing members (e.g., i-joists or trusses) deeper than 16 inches to be used if bearing wall stud heights were less than 10 feet. This was accomplished by specifying an overall story height limit of 11 feet 7 inches, based on the sum of a 10 feet 0 inch tall stud, two top and one bottom plate, and 16-inch deep floor framing. This limit superseded the exception which allowed bearing wall studs up to 12 feet tall with wall bracing per Section R602.10, also conflicting with the 12-foot bearing wall height limit for masonry walls.

In the 2015 IRC, this section was revised again by deleting the 11 feet 7 inches story height limit from Section R301.3 and placing it in each individual subsection to which it applied. This addressed the conflict with masonry walls but did not correct the conflict with Section R602.10. The exception for bearing wall studs up to 12 feet was deleted due to concern that code users would double-count adjustment factors for wall bracing. However, no correction was provided for the conflict between story height limits and wall bracing provisions.

A new Table R602.3(6) was added to the 2018 IRC allowing bearing wall stud heights up to 12 feet for limited conditions. The conflict between story height limits, wall bracing provisions and the new table was not addressed. The new provisions relied on the last paragraph of Section R301.3, which stated that individual walls or wall studs could exceed the limits of Section R301.3 when story heights were not exceeded.

The 2021 IRC restores the exception allowing bearing wall stud clear heights to be increased to 12 feet without engineering, provided compliance with Section R602.10 for wall bracing is met. Additionally, one of the two exceptions to 10-foot bearing wall heights in Section R602.3.1 is applicable, including the exception leading to Table R602.3(6). The story height exceptions provide a critical link to provisions allowing a stud height up to 12 feet without engineering. The engineering requirement for studs in tall walls not otherwise complying with one of the Section R602.3.1 exceptions is maintained for gravity and out-of-plane loads.

Section R602.3.1 limits studs above 10 feet in height to lower snow load regions. The ground snow load limit for Exception 2 is 25 pounds per square foot while the limit in Exception 3 is 30 pounds per square foot.

R302.2
Townhouses

CHANGE TYPE: Modification

CHANGE SUMMARY: Common walls separating townhouses are permitted to terminate at the inside of exterior walls where the prescribed fireblocking is provided.

2021 CODE: R302.2 Townhouses. Walls separating townhouse units shall be constructed in accordance with Section R302.2.1 or R302.2.2 <u>and shall comply with Sections 302.2.3 through 302.2.5</u>.

R302.2.1 Double walls. Each townhouse <u>unit</u> shall be separated <u>from other townhouse units</u> by two 1-hour fire-resistance-rated wall assemblies tested in accordance with ASTM E119, UL 263 or Section ~~703.3~~ <u>703.2.2</u> of the *International Building Code*.

R302.2.2 Common walls. Common walls separating ~~townhouses~~ <u>townhouse units</u> shall be assigned a fire-resistance rating in accordance with Item 1 or 2~~:~~ <u>and shall be rated for fire exposure from both sides. Common walls shall extend to and be tight against the exterior sheathing of the exterior walls, or the inside face of exterior walls without stud cavities, and the underside of the roof sheathing.</u> The common wall shared by two ~~townhouses~~ <u>townhouse units</u> shall be constructed without plumbing or mechanical equipment, ducts or vents<u>, other than water-filled fire sprinkler piping,</u> in the cavity of the common wall. ~~The wall shall be rated for fire exposure from both sides and shall extend to and be tight against exterior walls and the underside of the roof sheathing.~~ Electrical installations shall be in accordance with Chapters 34 through 43. Penetrations of the membrane of common walls for electrical outlet boxes shall be in accordance with Section R302.4.

1. Where ~~a fire~~ <u>an automatic</u> sprinkler system in accordance with Section P2904 is provided, the common wall shall be not less than a 1-hour fire-resistance- rated wall assembly tested in accordance with ASTM E119, UL 263 or Section ~~703.3~~ <u>703.2.2</u> of the *International Building Code*.

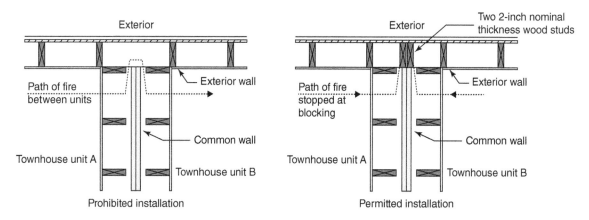

Common wall extending to the inside of the exterior wall.

2. Where ~~a fire~~ an automatic sprinkler system in accordance with Section P2904 is not provided, the common wall shall be not less than a 2-hour fire- resistance-rated wall assembly tested in accordance with ASTM E119, UL 263 or Section ~~703.3~~ 703.2.2 of the *International Building Code*.

   <u>**Exception:** Common walls are permitted to extend to and be tight against the inside of the exterior walls if the cavity between the end of the common wall and the exterior sheathing is filled with a minimum of two 2-inch nominal thickness wood studs.</u>

R302.2.3 Continuity. The fire-resistance-rated wall or assembly separating ~~townhouses~~ <u>townhouse units</u> shall be continuous from the foundation to the underside of the roof sheathing, deck or slab. The fire-resistance rating shall extend the full length of the wall or assembly, including wall extensions through and separating attached enclosed accessory structures.

R302.2.4 Parapets for townhouses. Parapets constructed in accordance with Section R302.2.5 shall be constructed for townhouses as an extension of exterior walls or common walls <u>separating townhouse units</u> in accordance with the following:

Items 1 through 3. *No changes to text.*

R302.2.6 Structural independence. Each ~~individual~~ townhouse <u>unit</u> shall be structurally independent.

   Exceptions:
   1. Foundations supporting exterior walls or common walls.
   2. Structural roof and wall sheathing from each unit fastened to the common wall framing.
   3. Nonstructural wall and roof coverings.
   4. Flashing at termination of roof covering over common wall.
   5. ~~Townhouses~~ <u>Townhouse units</u> separated by a common wall as provided in Section R302.2.2, Item 1 or 2.
   6. <u>Townhouse units protected by a fire sprinkler system complying with Section P2904 or NFPA 13D.</u>

CHANGE SIGNIFICANCE: Section R302.2 has been updated to incorporate the new definition of "townhouse unit" and the revised definition of "townhouse." The terms had sometimes been used interchangeably and inconsistently. In this context, a townhouse is a building that contains three or more dwelling units, now referred to as "townhouse units." The definition of building has also been updated to directly reference townhouses.

There are two recognized methods for the fire-resistant separation between townhouse units: either two 1-hour rated walls or a common wall with a 1-hour rating with fire sprinkler protection or a 2-hour rating without sprinklers. At issue has been the termination point of the common wall at the intersection with the exterior wall. The code language has been somewhat ambiguous in stating that the common wall must extend to and be tight against exterior walls. To some, this meant that the common wall had to extend through the wall to a termination point at the exterior wall sheathing, while others took it to mean a termination point at the interior side of the exterior wall. For typical frame cavity walls, the language has been revised to require the common wall to extend to and be tight against the exterior sheathing of the exterior wall to block the passage of fire from one townhouse unit to the next. A new exception allows the fire-resistance-rated wall to terminate against the interior side of the exterior wall provided at least two 2-inch nominal thickness wood studs fill the cavity in the exterior wall. Because the code language is very specific as to the fireblocking material, other types of fireblocking to fill this gap would not be allowed unless approved under the alternative methods provisions in Section R104.11.

Where two 1-hour fire-resistance-rated walls are used as the separation, structural independence related to fire resistance is required for each townhouse unit. Some exceptions to this rule appear in Section R302.2.6, including the application of exterior sheathing and finish materials. A new exception strikes the structural independence rule if the townhouse units are protected with a fire sprinkler system complying with Section P2904 or NFPA 13D. This adds another sprinkler incentive to the code to elevate the safety of IRC buildings in those areas of the country that do not adopt the sprinkler provisions of the model code, which require sprinklers in all dwellings and townhouses.

Common walls for separating townhouse units enjoy an advantage in that structural independence is not required. However, the code does add some other conditions regarding common walls; specifically, plumbing and mechanical piping, equipment, ducts and vents are not allowed in the common wall. In a new exception, the code now permits water filled sprinkler piping to be in the common wall based on the added safety provided by sprinklers outweighing any negative impact on the effectiveness of the rated common wall. Allowing common fire sprinkler piping to protect multiple townhouse units in a townhouse building can significantly reduce installation costs, and the provision is consistent with the IBC, which allows penetration of townhouse separation walls in any townhouse that does not exceed the height and area limits. See Sections R302.4.1 and R302.4.2 for additional information on sprinkler piping penetrations of fire-resistance-rated assemblies.

R302.3
Two-Family Dwelling Separation

CHANGE TYPE: Modification

CHANGE SUMMARY: The prescribed fire-resistance-rated separation between two dwelling units in a single building is not affected by the presence of a lot line between the units.

2021 CODE: R302.3 Two-family dwellings. Dwelling units in two-family dwellings shall be separated from each other by wall and floor assemblies having not less than a 1-hour fire-resistance rating where tested in accordance with ASTM E119, UL 263 or Section ~~703.3~~ 703.2.2 of the *International Building Code*. <u>Such separation shall be provided regardless of whether a lot line exists between the two dwelling units or not.</u> Fire-resistance-rated floor/ceiling and wall assemblies shall extend to and be tight against the exterior wall, and wall assemblies shall extend from the foundation to the underside of the roof sheathing.

Exceptions:
1. A fire-resistance rating of 1/2 hour shall be permitted in buildings equipped throughout with an automatic sprinkler system installed in accordance with ~~NFPA 13~~ <u>Section P2904</u>.
2. Wall assemblies need not extend through attic spaces where the ceiling is protected by not less than 5/8-inch (15.9 mm) Type X gypsum board, an attic draft stop constructed as specified in Section R302.12.1 is provided above and along the wall assembly separating the dwellings and the structural framing supporting the ceiling is protected by not less than 1/2- inch (12.7 mm) gypsum board or equivalent.

Two-family dwelling.

CHANGE SIGNIFICANCE: Unlike townhouse unit separations, two-family dwellings (duplexes) only require a 1-hour fire-resistance-rated separation between dwelling units. It has been debated whether a lot line between

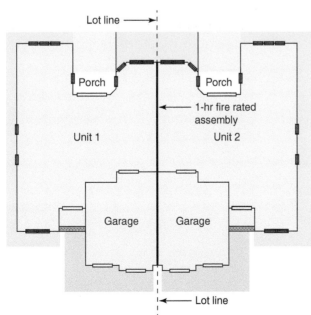

Two-family dwelling separated by lot line and 1-hr fire-resistant separation.

the dwelling units (which is common in some areas of the country and not common in others) impacts this separation requirement. The question has been whether the lot line means that the wall at the separation is considered to be an exterior wall that needs to meet the provisions of Section R302.1, resulting in two 1-hour walls at the lot line. In some jurisdictions, the answer was yes. Further, in some cases with a separating lot line, the interpretation has been that the building is no longer a two-family dwelling, but two separate detached single-family dwellings, each requiring a 1-hour wall at the lot line. In other jurisdictions, the answer was no: a duplex is a duplex no matter if the dwelling units are divided by a lot line. The reasoning behind this approach was that the fire does not know if there is a lot line there and only the 1-hour separation applies. The change to this section intends to end the debate and clarify the application of this separation. The intent of the new language is that a fire-resistance rating need never be greater than 1 hour, whether there is a lot line between dwelling units or not. For the lot line question, this brings the two-family dwelling provisions into agreement with the townhouse provisions. If the townhouse has fire sprinkler protection, a common 1-hour wall has been acceptable even if there was a lot line between townhouse units. If the exterior wall provisions in Section R302.1 were applied to townhouses, the 1-hour common wall would not be allowed.

Presumably, this change to the code allowing a 1-hour separation when there is a lot line between duplex dwelling units is meant to apply to the exception as well. The exception permits a draft stop to separate the attics of the dwelling units if other fire-resistance requirements are satisfied.

Another exception to the 1-hour separation requirement for two-family dwellings has allowed the rating to be reduced to ½ hour if a full NFPA 13 sprinkler system was installed. This exception has not been used nor would it be used because of the extra cost associated with a full NFPA 13 system typically associated with commercial structures. The cost would far outweigh any savings realized from reducing the rating from 1 hour to ½ hour. As another incentive to install a sprinkler system for areas of the country that do not adopt the IRC sprinkler provisions, a dwelling sprinkler system installed in accordance with Section P2904 or NFPA 13D now can be used to reduce the rating to ½ hour.

R302.4
Dwelling Unit Rated Penetrations

CHANGE TYPE: Clarification

CHANGE SUMMARY: Water-filled fire sprinkler piping of any approved material joins the list of metal penetrating items that do not require a firestop system provided the annular space is filled with the prescribed materials.

2021 CODE: R302.4.1 Through penetrations. Through penetrations of fire-resistance-rated wall or floor assemblies shall comply with Section R302.4.1.1 or R302.4.1.2.

Exceptions:

1. Where the penetrating items are steel, ferrous or copper pipes, tubes or conduits, the annular space shall be protected as follows:

    ~~1.~~ 1.1 In concrete or masonry wall or floor assemblies, concrete, grout or mortar shall be permitted where installed to the full thickness of the wall or floor assembly or the thickness required to maintain the fire-resistance rating, provided that both of the following are complied with:

    ~~1.1~~ 1.1.1. The nominal diameter of the penetrating item is not more than 6 inches (152 mm).

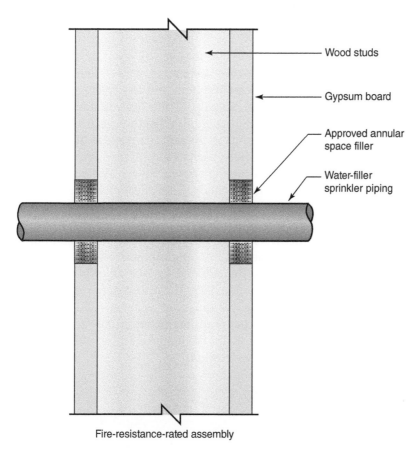

Water-filled fire sprinkler piping penetrating a fire-rated assembly.

~~1.2~~ 1.1.2. The area of the opening through the wall does not exceed 144 square inches (92 900 mm).

~~2.~~ 1.2. The material used to fill the annular space shall prevent the passage of flame and hot gases sufficient to ignite cotton waste where subjected to ASTM E119 or UL 263 time temperature fire conditions under a positive pressure differential of not less than 0.01 inch of water (3 Pa) at the location of the penetration for the time period equivalent to the fire-resistance rating of the construction penetrated.

2. <u>The annular space created by the penetration of water-filled fire sprinkler piping, provided that the annular space is filled using a material complying with item 1.2 of Exception 1.</u>

R302.4.2 Membrane penetrations. Membrane penetrations shall comply with Section R302.4.1. Where walls are required to have a fire-resistance rating, recessed fixtures shall be installed so that the required fire-resistance rating will not be reduced.

Exceptions:

1. **and 2.** *No changes to text.*
3. The annular space created by the penetration of a fire sprinkler <u>or water-filled fire sprinkler piping,</u> provided that ~~it~~ <u>the annular space</u> is covered by a metal escutcheon plate.
4. *No changes to text.*

CHANGE SIGNIFICANCE: When items such as pipes or ducts penetrate one or both sides of the fire-resistance-rated wall assembly separating dwelling units, both the penetrating item and the space around it must be protected to maintain the integrity of the fire-resistant assembly. In general, penetrations by metal pipe require that the space around the pipe be filled with approved materials to prevent the passage of flame and hot gases. Other penetrating materials, such as plastic pipe, must be protected by an approved penetration fire-stop system. Such a system often consists of intumescent material that expands when heated by fire conditions, filling the penetration as the plastic pipe melts and preserving the fire-resistance rating of the wall assembly.

Listed nonmetallic fire sprinkler piping is ignition resistant and will not sustain combustion. The IRC now permits water-filled fire sprinkler piping to penetrate a fire-resistance-rated membrane or both membranes of a through penetration without a listed firestop system provided that the annular space is filled using an approved material. This matches the installation requirements for metal piping penetrations. Exception 1.2 of Section R302.4.1 sets the criteria for the material filling the annular space of the membrane or through penetration.

R302.5 Dwelling-Garage Opening Protection

CHANGE TYPE: Clarification

CHANGE SUMMARY: Doors between the garage and residence must be self-latching.

2021 CODE: R302.5 Dwelling-garage opening and penetration protection. Openings and penetrations through the walls or ceilings separating the dwelling from the garage shall be in accordance with Sections R302.5.1 through R302.5.3.

R302.5.1 Opening protection. Openings from a private garage directly into a room used for sleeping purposes shall not be permitted. Other openings between the garage and residence shall be equipped with solid wood doors not less than 1 3/8 inches (35 mm) in thickness, solid or honeycomb-core steel doors not less than 1 3/8 inches (35 mm) thick, or 20-minute fire-rated doors. <u>Doors shall be self-latching and</u> equipped with a self-closing or an automatic-closing device.

CHANGE SIGNIFICANCE: To provide some minimum protection against the spread of a fire that originates in the attached garage, the IRC has always required some fire resistance for the separation between the garage and dwelling unit. Typically, this requirement is satisfied with the application of regular 1/2-inch gypsum board on the garage side of the separation. This separation is not a fire-resistant-rated assembly, but simply a layer of approved material installed on the garage side to provide some resistance to fire that originates in the garage and slows spread into the dwelling unit. Similarly, the code does not require a fire-resistant-rated door assembly for the opening between the garage and residence. Instead, the IRC prescribes the type and thickness of the door or requires a 20-minute rating for the door slab. The requirement for self-closing devices introduced in the 2012 IRC intended that the door

Door from garage to house must be self-closing and self-latching.

return to a closed position after opening to address concerns related to increased fuel loads in garages, and the potential for fire and the related toxic combustion byproducts to migrate into the dwelling unit. Although Sections R302.5 and R302.6 are primarily concerned with fire resistance, the decision to place self-closing devices in the code was also intended to prevent carbon monoxide from the exhaust of vehicles operating in a garage from entering the dwelling unit. Having a closed door between the garage and residence supplements the safeguards of required smoke and carbon monoxide alarms. Self-closing devices are typically spring-loaded hinges or door closers. An automatic closing device is also permitted but is not typically seen in one- and two- family dwellings.

New to the 2021 edition, the code adds that the self-closing door must also be self-latching. In this case the smooth operation of the door and strength of the self-closing device need to be sufficient for the latch to engage and the door to be secured. Self-latching was judged to be as important as self-closing. Air pressures caused by a fire could cause doors without a self-latching device to overpower the closing device and push open. The provision for dwelling-garage opening protection is consistent with the premise that closed doors limit the spread and impact of residential fires and addresses the increased fire hazard in garages.

CHANGE TYPE: Clarification

CHANGE SUMMARY: A local exhaust system is an acceptable substitute for natural ventilation in kitchens.

2021 CODE: R303.1 Habitable rooms. Habitable rooms shall have an aggregate glazing area of not less than 8 percent of the floor area of such rooms. Natural ventilation shall be through windows, skylights, doors, louvers or other approved openings to the outdoor air. Such openings shall be provided with ready access or shall otherwise be readily controllable by the building occupants. The openable area to the outdoors shall be not less than 4 percent of the floor area being ventilated.

Exceptions:

1. ~~The~~ <u>For habitable rooms other than kitchens, the</u> glazed areas need not be openable where the opening is not required by Section R310 and a whole-house mechanical ventilation system <u>or a mechanical ventilation system capable of producing 0.35 air changes per hour in the habitable rooms</u> is installed in accordance with Section M1505.

2. <u>For kitchens, the glazed areas need not be openable where the opening is not required by Section R310 and a local exhaust system is installed in accordance with Section M1505.</u>

~~2.~~<u>3.</u> The glazed areas need not be installed in rooms where Exception 1 is satisfied and artificial light is provided that is capable of producing an average illumination of 6 footcandles (65 lux) over the area of the room at a height of 30 inches (762 mm) above the floor level.

R303.1
Mechanical Ventilation

Local exhaust mechanical ventilation for a kitchen substitutes for an openable window.

~~3.~~ **4.** Use of sunroom and patio covers, as defined in Section R202, shall be permitted for natural ventilation if in excess of 40 percent of the exterior sunroom walls are open, or are enclosed only by insect screening.

R303.4 Mechanical ventilation. ~~Where the air infiltration rate of a dwelling unit is 5 air changes per hour or less where tested with a blower door at a pressure of 0.2 inch w.c. (50 Pa) in accordance with Section N1102.4.1.2, the dwelling unit~~ <u>Buildings and dwelling units complying with Section N1102.4.1</u> shall be provided with ~~whole-house~~ mechanical ventilation in accordance with Section M1505~~.4~~, <u>or with other approved means of ventilation</u>.

CHANGE SIGNIFICANCE: To provide flexibility in satisfying the ventilation air requirements, the code adds some options for mechanical ventilation. Prior to the 2012 edition, the IRC provided an option for mechanical ventilation systems capable of producing 0.35 air changes per hour (ACH) or whole-house mechanical ventilation systems where the prescribed natural ventilation was not provided. In the 2012 IRC, the first option was deleted in favor of the whole house mechanical ventilation system covered in the mechanical provisions of Chapter 15. The 2021 code reintroduces the option for 0.35 ACH for room-based mechanical ventilation systems, which are especially useful for additions or remodels. The code also now specifically recognizes that a local exhaust system (e.g., a kitchen exhaust hood) is an acceptable substitute for natural ventilation in kitchens. This is consistent with provisions in Chapter 15 that recognize kitchens have different ventilation requirements than other habitable rooms. The intent is to clarify that local exhaust, not whole-house mechanical ventilation that could be located in a far corner of the house, is needed for kitchens to ensure that cooking pollutants generated in the kitchen are captured and exhausted at their source. The IRC sets installation requirements for kitchen exhaust but has not previously required a range hood. The intent of the change to the exceptions in Section R303.1 is to require local mechanical exhaust in the kitchen if natural ventilation through openable windows and doors is not provided.

R305.1 ■ Ceiling Height

R305.1
Ceiling Height

CHANGE TYPE: Modification

CHANGE SUMMARY: The minimum ceiling height is reduced to 6 feet 6 inches under beams spaced at least 36 inches apart.

2021 CODE: R305.1 Minimum height. Habitable space, hallways and portions of basements containing these spaces shall have a ceiling height of not less than 7 feet (2134 mm). Bathrooms, toilet rooms and laundry rooms shall have a ceiling height of not less than 6 feet 8 inches (2032 mm).

Exceptions:

1. For rooms with sloped ceilings, the required floor area of the room shall have a ceiling height of not less than 5 feet (1524 mm) and not less than 50 percent of the required floor area shall have a ceiling height of not less than 7 feet (2134 mm).
2. The ceiling height above bathroom and toilet room fixtures shall be such that the fixture is capable of being used for its intended purpose. A shower or tub equipped with a showerhead shall have a ceiling height of not less than 6 feet 8 inches (2032 mm) above an area of not less than 30 inches (762 mm) by 30 inches (762 mm) at the showerhead.
3. Beams, girders, ducts or other obstructions in basements containing habitable space shall be permitted to project to within 6 feet 4 inches (1931 mm) of the finished floor.
4. <u>Beams and girders spaced apart not less than 36 inches (914 mm) in clear finished width shall project not more than 78 inches (1981 mm) from the finished floor.</u>

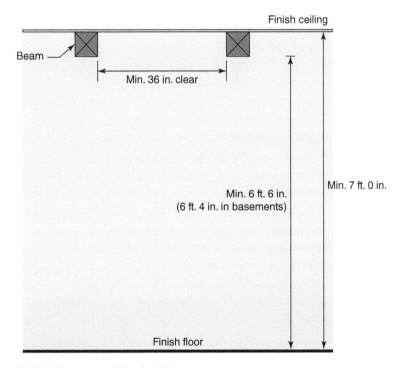

Habitable space ceiling heights.

CHANGE SIGNIFICANCE: Prior to the 2009 edition of the IRC, an exception to the general rule for a minimum ceiling height of 7 feet allowed beams to project 6 inches below that ceiling height provided they were spaced at least 4 feet apart. That language was removed in the 2009 code because there was consensus that the minimum ceiling height of 7 feet was reasonable to be maintained under spaced beams. At that time, portions of unfinished basements could have beams, girders and ducts with a minimum ceiling height of 6 feet 4 inches. In the 2015 IRC, that exception was expanded to include these projections in all spaces in basements including habitable space.

Although the new exception sets a maximum projection from the floor, the intent of the language in the code is to allow beams to project from the ceiling such that the base of the beam is at least 6 feet 6 inches above the floor. The minimum finish-to-finish horizontal dimension between the beams is 36 inches. The exception would not be applicable to basements as the code already permits a ceiling height reduction to 6 feet 4 inches for beams, girders and ductwork in basements. The new exception is a result of concern expressed that the size of beams and girders has increased in recent years with the advent of engineered lumber and open floor plans, and that a ceiling height of 7 feet was not always achievable. The height of 6 feet 6 inches was chosen based on the clear height dimension of the one required egress door measured from the threshold to the underside of the head stop. The code does not regulate the dimensions of other doors in the house, but typically they maintain a clear height exceeding 6 feet 6 inches.

R308.4.5
Glazing and Wet Surfaces

CHANGE TYPE: Clarification

CHANGE SUMMARY: The language addressing glazing in walls, enclosures or fences near tubs, showers and swimming pools has replaced the word "facing" with the words "adjacent to" for those elements related to wet surfaces.

2021 CODE: R308.4.5 Glazing and wet surfaces. Glazing in walls, enclosures or fences containing or ~~facing~~ <u>adjacent to</u> hot tubs, spas, whirlpools, saunas, steam rooms, bathtubs, showers and indoor or outdoor swimming pools where the bottom exposed edge of the glazing is less than 60 inches (1524 mm) measured vertically above any standing or walking surface shall be considered to be a hazardous location. This shall apply to single glazing and each pane in multiple glazing.

Exception: Glazing that is more than 60 inches (1524 mm), measured horizontally ~~and in a straight line~~, from the water's edge of a bathtub, hot tub, spa, whirlpool or swimming pool or from the edge of a shower, sauna or steam room.

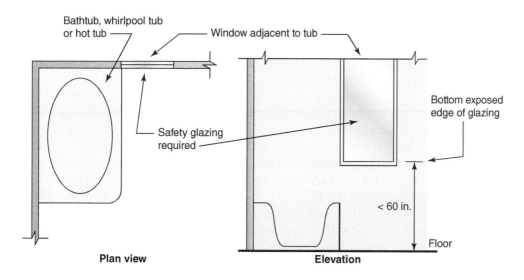

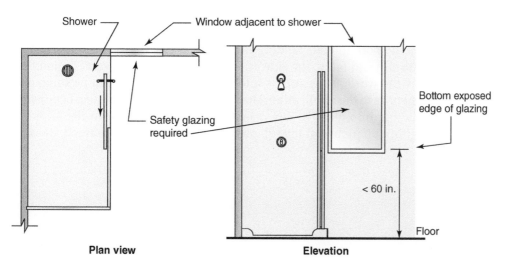

Glazing adjacent tubs and showers.

CHANGE SIGNIFICANCE: Glazing installed in the vicinity of tubs, showers and swimming pools is considered to be in a hazardous location because of the increased likelihood of slips and falls on wet surfaces, with the potential for injury from falling into and breaking the glass. Therefore, the code requires that glazing installed in these locations must be safety glazing. As with all of the hazardous locations identified in Section R308.4, changes to the language in this section over several code development cycles have attempted to make the provisions more objectively measurable and easier to understand.

There has been some debate as to the proper interpretation and application of the glazing provisions related to wet surfaces. It has been clear that glass enclosing tubs, showers, spas and swimming pools (to name the common items under consideration) has to be safety glazing unless meeting the minimum height above the walking surface. Less clear is the application to glazing that is located some distance horizontally from these fixtures and areas. The code has generally established a zone that is within 60 inches horizontally of the water's edge (the edge of a shower) as a potentially wet surface and hazardous location. The 60-inch measurement originated in the early editions of the code where it was applied to swimming pools to create a 60-inch buffer zone between the water and standard glazing. The 60-inch horizontal measurement relating to tubs and showers first appeared in the 2009 edition of the IRC. The safety glazing rules for swimming pools and tubs and showers were merged in the 2012 edition of the code. Part of the ongoing confusion is that, at least since the 2012 edition, the 60-inch horizontal measurement is written as an exception to the main rule.

It follows that various interpretations are not always in agreement. For example, a majority of code users have determined that glazing within 60 inches in any direction requires safety glazing, unless there is some barrier in place to separate the wet surface from the glazing. Others have said that only the glazing that faces the tub, shower or swimming pool is regulated in that 60-inch zone and exclude glazing that is perpendicular and not enclosing the fixture. This change intends to remedy these differences by using the word "adjacent" rather than "facing" to describe the location of the glazing. This is similar to language used to describe glazing adjacent to the bottom stair landing. For stairs, the term adjacent and the 60-inch dimension appear together in the main code section with text accompanied by a figure to clarify the application.

The intent of the change to Section R308.4.5, Glazing and Wet Surfaces, is to apply the horizontal measurement provision in the same way as the bottom stair landing in Section R308.4.7. A person is just as likely to slip and fall to the side as a a person is to fall forward. Depending on how the word "adjacent" is interpreted, there may again be some disagreement as to the application. However, the intent is to protect occupants from falling into glass in any orientation when they are in the slippery area within 60 inches of the tub or shower or the other described areas.

R308.6
Skylight Glass Retention Screens

CHANGE TYPE: Clarification

CHANGE SUMMARY: New terminology clarifies the broken glass retention screen requirements for skylights.

2021 CODE: R308.6.2 Materials. Glazing materials shall be limited to the following:

1. Laminated glass with not less than a 0.015-inch (0.38 mm) polyvinyl butyral interlayer for glass panes 16 square feet (1.5 m²) or less in area located such that the highest point of the glass is not more than 12 feet (3658 mm) above a walking surface; for higher or larger sizes, the interlayer thickness shall be not less than 0.030 inch (0.76 mm).
2. Fully tempered glass.
3. Heat-strengthened glass.
4. Wired glass.
5. Approved rigid plastics.

R308.6.3 Screens, general. For fully tempered or heat-strengthened glass, a ~~retaining~~ <u>broken glass retention</u> screen meeting the requirements of Section R308.6.7 shall be installed below <u>the full area of</u> the glass, except for fully tempered glass that meets ~~either~~ condition <u>1 or 2</u> listed in Section R308.6.5.

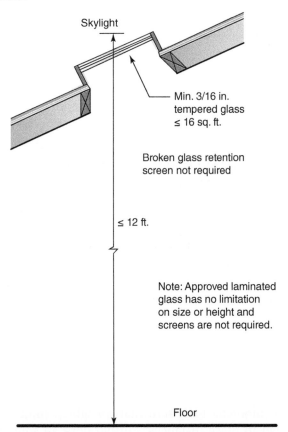

Broken glass retention screens are not required for laminated glass or when meeting the size, height and material limitations for tempered glass.

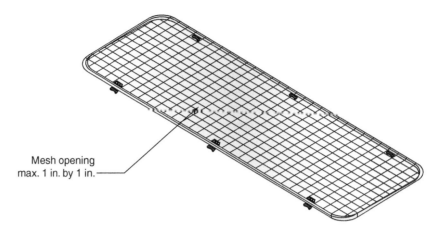

Broken glass retention screen.

R308.6.4 Screens with multiple glazing. Where the inboard pane is fully tempered, heat-strengthened or wired glass, a ~~retaining~~ broken glass retention screen meeting the requirements of Section R308.6.7 shall be installed below the full area of the glass, except for ~~either condition~~ Condition 1 or 2 listed in Section R308.6.5. Other panes in the multiple glazing shall be of any type listed in Section R308.6.2.

R308.6.5 Screens not required. Screens shall not be required where laminated glass complying with item 1 of Section R308.6.2 is used as single glazing or the inboard pane in multiple glazing. Screens shall not be required where fully tempered glass is used as single glazing or the inboard pane in multiple glazing and either of the following conditions are met:

1. The glass area is 16 square feet (1.49 m^2) or less; the highest point of glass is not more than 12 feet (3658 mm) above a walking surface; the nominal glass thickness is not more than 3/16 inch (4.8 mm); and, ~~(for multiple glazing only)~~, the other pane or panes are fully tempered, laminated or wired glass.
2. The glass area is greater than 16 square feet (1.49 m^2); the glass is sloped 30 degrees (0.52 rad) or less from vertical; and the highest point of glass is not more than 10 feet (3048 mm) above a walking surface.

R308.6.7 Screen characteristics. The screen and its fastenings shall: be capable of supporting twice the weight of the glazing~~;~~ ; be firmly and substantially fastened to the framing members~~,~~; be installed within 4 inches (102mm) of the glass; and have a mesh opening of not ~~more~~ greater than 1 inch by 1 inch (25 mm by 25 mm).

CHANGE SIGNIFICANCE: Previous language related to the installation of "retaining" screens below unit skylights has caused some confusion and occasionally misinterpretation resulting in the installation of screens where they were not required. In some cases, skylight manufacturers have been asked to intervene to ensure that unsightly, unnecessary screens are not installed in these instances. Furthermore, in some instances, an optional skylight installation is removed from submitted plans due to misinterpretation at the plan review stage. In reality, glass retention screens are rarely required in IRC buildings and are never required if qualifying laminated glass is used in the skylight. The previous code language addressed qualifying laminated glass by simple omission from the sections dealing with screens. It was this omission that seemed to create some confusion within the industry. The added sentence in Section R308.6.5 states directly that permitted laminated glass does not require retention screens.

To further clarify the application, the terminology has been modified to fully describe the purpose of the screens as broken glass retention screens. This is to ensure readers do not confuse them with insect screens or fall protection screens, which are physically different and will not serve as effective retention screens. The hazard being addressed is to retain larger pieces of broken glass from falling on and injuring occupants.

Changes to Section R308.6.7 addressing the physical characteristics of the retention screens, if required, are consistent with the language in IBC Section 2405.3. None of the changes to the code text affect the long-standing requirements, but simply clarify the provisions to ensure more consistent enforcement.

R310.1

Emergency Escape and Rescue Opening Required

CHANGE TYPE: Clarification

CHANGE SUMMARY: Emergency escape and rescue openings require a clear 36-inch-wide path to a public way. Operation requirements have been clarified.

2021 CODE: R310.1 Emergency escape and rescue opening required. Basements, habitable attics and every sleeping room shall have not less than one operable emergency escape and rescue opening. Where basements contain one or more sleeping rooms, an emergency escape and rescue opening shall be required in each sleeping room. Emergency escape and rescue openings shall open directly into a public way, or to a yard or court <u>having a minimum width of 36 inches (914 mm)</u> that opens to a public way.

Exceptions:

1. Storm shelters and basements used only to house mechanical equipment not exceeding a total floor area of 200 square feet (18.58 m²).

2. Where the dwelling <u>unit</u> or townhouse <u>unit</u> is equipped with an automatic sprinkler system installed in accordance with Section P2904, sleeping rooms in basements shall not be required to have emergency escape and rescue openings provided that the basement has one of the following:

    2.1. One means of egress complying with Section R311 and one emergency escape and rescue opening.

    2.2. Two means of egress complying with Section R311.

3. <u>A yard shall not be required to open directly into a public way where the yard opens to an unobstructed path from the yard to the public way. Such path shall have a width of not less than 36 inches (914 mm).</u>

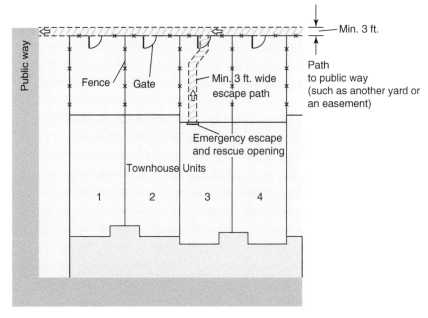

Min. 3-ft. wide path required from an emergency escape and rescue opening to the public way.

R310.1.1 Operational constraints and opening control devices.
Emergency escape and rescue openings shall be operational from the inside of the room without the use of keys, tools or special knowledge. Window opening control devices <u>and fall prevention devices complying with ASTM F2090 shall be permitted for use</u> on windows serving as a required emergency escape and rescue opening <u>and shall be not more than 70 inches (178 cm) above the finished floor</u> ~~shall comply with ASTM F2090~~.

CHANGE SIGNIFICANCE: Emergency escape and rescue openings have always been one of the most important fire- and life-safety provisions of the code. Although the provisions have been revised each code cycle, the basic concept to provide a secondary means out of a building in an emergency have held steady since the first edition of the IRC in 2000.

The openings have always been required in sleeping rooms. Originally required in basements with habitable space, beginning with the 2006 code they have been required in all basements, with some minor exceptions based on use. The 2006 edition also introduced the requirement to provide direct access from an emergency escape and rescue opening to the public way or to a yard or court that opens to a public way. The intent has been to connect directly to the outdoors so that travel through another room or inside space is not necessary. The code has not previously prescribed the width of the path of travel, instead relying on more of a performance provision that a path be adequate to travel to a public way. New language in the 2021 code requires a path not less than 36 inches in width for travel to the public way. The code does stipulate that the yard is not required to open directly into a public way provided it opens to an unobstructed path (typically another yard) that leads to the public way. The new path width requirement also applies to emergency escape and rescue openings under decks, porches and cantilevers as discussed in Section R310.2 in this publication.

The code clarifies that window opening control devices and fall prevention devices in compliance with ASTM F2090, *Standard Specification for Window Fall Prevention Devices with Emergency Escape (Egress) Release Mechanisms* are approved for use on emergency escape and rescue windows. New text limits the height of the release mechanisms to 70 inches above the finished floor. These devices are used to prevent children from falling out of windows where the bottom of the opening exceeds 6 feet above the ground outside, as covered in Section R312. Provisions are placed in this section as a reminder to code users that these devices are acceptable for escape and rescue applications.

R310.2
Emergency Escape and Rescue Openings

CHANGE TYPE: Modification

CHANGE SUMMARY: Emergency escape openings under decks, porches and cantilevers require a path not less than 36 inches wide. Opening dimensions have been clarified.

2021 CODE: R310.2 Emergency escape and rescue openings. Emergency escape and rescue openings shall have minimum dimensions ~~as specified in this section~~ <u>in accordance with Sections R310.2.1 through R310.2.4</u>.

R310.2.1 Minimum ~~opening area~~ <u>size</u>. Emergency and escape rescue openings shall have a net clear opening of not less than 5.7 square feet (0.530 m^2). ~~The net clear opening dimensions required by this section shall be obtained by the normal operation of the emergency escape and rescue opening from the inside. The net clear height of the opening shall be not less than 24 inches (610 mm) and the net clear width shall be not less than 20 inches (508 mm).~~

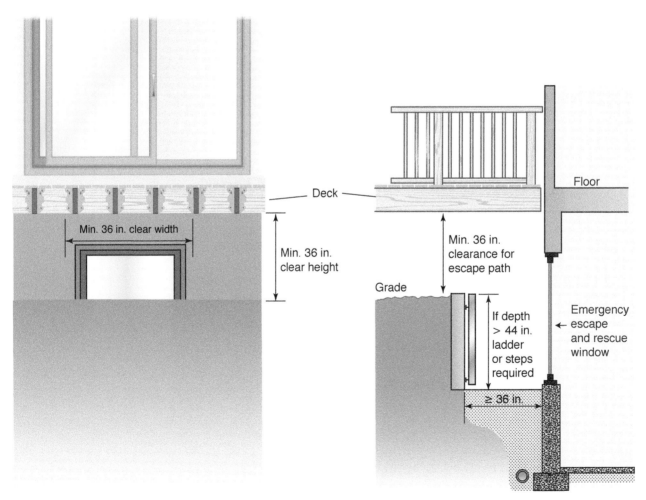

Section at deck facing houseCross section of area well below deck

Minimum width and height for emergency escape path below a deck or porch.

Exception: ~~Grade floor openings or below-grade openings shall have a net clear opening area of not less than~~ The minimum net clear opening for grade-floor emergency escape and rescue openings shall be 5 square feet (0.465 m²).

R310.2.2 Minimum dimensions. The minimum net clear opening height dimension shall be 24 inches (610 mm). The minimum net clear opening width dimension shall be 20 inches (508 mm). The net clear opening dimensions shall be the result of normal operation of the opening.

~~R310.2.2~~ R310.2.3 ~~Window sill height~~ Maximum height from floor. ~~Where a window is provided as the emergency~~ Emergency escape and rescue ~~opening,~~ openings ~~it~~ shall have ~~a sill height of not more than~~ the bottom of the clear opening not greater than 44 inches (1118 mm) above the floor.~~; where the sill height is below grade, it shall be provided with a window well in accordance with Section R310.2.3.~~

~~R310.2.4~~ R310.2.4 Emergency escape and rescue openings under decks ~~and~~, porches and cantilevers. Emergency escape and rescue openings installed under decks ~~and~~, porches and cantilevers shall be fully openable and provide a path not less than 36 inches (914 mm) in height and 36 inches (914 mm) in width to a yard or court.

CHANGE SIGNIFICANCE: Most of the changes to Section R310.2, with a couple of exceptions, are editorial and organizational in nature.

The minimum net clear opening area (5.7 square feet) and the minimum opening dimensions (20 inches in width and 24 inches in height) are placed in separate sections. The exception for a grade floor opening allowing a minimum area of 5.0 square feet has been clarified and matches the revised definition, now titled "Grade Floor Emergency Escape and Rescue Opening." The code previously stated that the exception applied to grade floor openings and below grade openings, not recognizing that "grade floor opening" was a defined term in the code. There was some question as to whether a "below grade opening" could be deeper than 44 inches below grade (a dimension from the definition of grade floor opening) and still take advantage of the size reduction. To make the connection clear, the definition has been revised and the exception in Section R310.2.1 has been placed in the section text and applies only to an opening meeting the defined term "Grade Floor Emergency Escape and Rescue Opening."

Consistent with several other changes (e.g., Definitions and Sections R310 and R312) in the 2021 IRC related to measurements from the floor to window openings, references to sills and window sills have been removed in favor of dimensions measured to the clear opening of the window. Such is the case with Section R310.2.4 where the title has been changed from "Window sill height" to "Maximum height from floor." The revised text clarifies that the measurement for the maximum height of 44 inches for an emergency escape window is from the finished floor to the bottom of the clear opening. Interestingly, this measurement language appeared in the 2012 IRC but was inadvertently dropped from the 2015 and 2018 editions.

The code has specifically permitted emergency escape and rescue openings under decks and porches in previous editions. A minimum height of 36 inches has been required for the path under the deck or porch to the yard. New to the 2021 IRC, openings under floor cantilevers are also allowed and the path to the yard must have a minimum width of 36 inches in addition to the 36-inch height requirement.

R310.3, R310.4

Area Wells for Emergency Escape and Rescue Openings

CHANGE TYPE: Modification

CHANGE SUMMARY: The provisions for window wells and area wells serving emergency escape and rescue openings have been merged into one section for area wells.

2021 CODE: R310.3 Emergency escape and rescue doors. Where a door is provided as the required emergency escape and rescue opening, it shall be a side-hinged door or a ~~slider. Where the opening is below the adjacent grade, it shall be provided with an area well.~~ sliding door.

~~**R310.3.1 Minimum door opening size.** The minimum net clear height opening for any door that serves as an emergency and escape rescue opening shall be in accordance with Section R310.2.1.~~

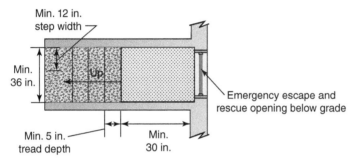

Plan view

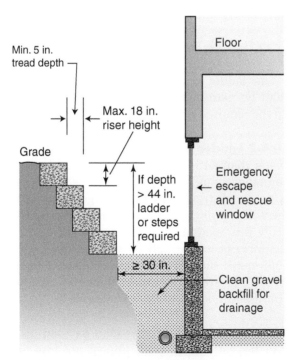

Area well with steps

Section

Step dimensions for area wells exceeding 44 inches in depth.

R310.3.2 Area wells. ~~Area wells shall have a width of not less than 36 inches (914 mm). The area well shall be sized to allow the emergency escape and rescue door to be fully opened.~~

R310.3.2.1 Ladder and steps. ~~Area wells with a vertical depth greater than 44 inches (1118 mm) shall be equipped with a permanently affixed ladder or steps usable with the door in the fully open position. Ladders or steps required by this section shall not be required to comply with Section R311.7. Ladders or rungs shall have an inside width of not less than 12 inches (305 mm), shall project not less than 3 inches (76 mm) from the wall~~ and shall be spaced not more than 18 inches (457 mm) on center vertically for the full height of the exterior stairwell.

R310.3.2.2 Drainage. ~~Area wells shall be designed for proper drainage by connecting to the building's foundation drainage system required by Section R405.1 or by an approved alternative method.~~

> **Exception:** ~~A drainage system for area wells is not required where the foundation is on well-drained soil or sand-gravel mixture soils in accordance with the United Soil Classification System, Group I Soils, as detailed in Table R405.1.~~

R310.4 Area wells. <u>An emergency escape and rescue opening where the bottom of the clear opening is below the adjacent grade shall be provided with an area well in accordance with Sections R310.4.1 through R310.4.4.</u>

~~R310.2.3~~ <u>R310.4.1</u> ~~Window wells.~~ <u>Minimum size.</u> The horizontal area of the ~~window~~ <u>area</u> well shall be not less than 9 square feet (0.9 m²), with a horizontal projection and width of not less than 36 inches (914 mm). The ~~area~~ <u>size</u> of the ~~window~~ <u>area</u> well shall allow the emergency escape and rescue opening to be fully opened.

> **Exception:** The ladder or steps required by Section ~~R310.2.3.1~~ <u>R310.2.4.2</u> shall be permitted to encroach not more than 6 inches (152 mm) into the required dimensions of the ~~window~~ <u>area</u> well.

~~R310.2.3.1~~ <u>R310.4.2</u> Ladder and steps. ~~Window~~ <u>Area</u> wells with a vertical depth greater than 44 inches (1118 mm) shall be equipped with ~~a~~ <u>an approved</u> permanently affixed <u>ladder or steps. The</u> ladder or steps ~~usable with~~ <u>shall not be obstructed by the emergency escape and rescue opening where</u> the window <u>or door is</u> in the ~~fully~~ open position. Ladders or steps required by this section shall not be required to comply with Section R311.7. ~~Ladders or rungs shall have an inside width of not less than 12 inches (305 mm), shall project not less than 3 inches (76 mm) from the wall and shall be spaced not more than 18 inches (457 mm) on center vertically for the full height of the window well.~~

R310.4.2.1 Ladders. <u>Ladders and rungs shall have an inside width of not less than 12 inches (305 mm), shall project not less than 3 inches (76 mm) from the wall and shall be spaced not more than 18 inches (457 mm) on center vertically for the full height of the area well.</u>

R310.4.2.2 Steps. Steps shall have an inside width of at least 12 inches (305 mm), a minimum tread depth of 5 inches (127 mm) and a maximum riser height of 18 inches (457 mm) for the full height of the area well.

R310.2.3.2 R310.4.3 Drainage. ~~Window~~ Area wells shall be designed for proper drainage by connecting to the building's foundation drainage system required by Section R405.1 ~~or by an approved alternative method~~.

> **Exception:** A drainage system for ~~window~~ area wells is not required where the foundation is on well-drained soil or sand-gravel mixture soils in accordance with the United Soil Classification System, Group I Soils, as detailed in Table R405.1.

R310.4 R310.4.4 Bars, grilles, covers and screens. Where bars, grilles, covers, screens or similar devices are placed over emergency escape and rescue openings, ~~area wells~~ bulkhead enclosures, or ~~window~~ area wells that serve such openings, the minimum net clear opening size shall comply with Sections ~~R310.2.1~~ R310.2 through ~~R310.2.3,~~ R310.2.2 and ~~such~~ R310.4.1. Such devices shall be releasable or removable from the inside without the use of a key, or tool, ~~special knowledge~~ or force greater than that required for the normal operation of the escape and rescue opening.

CHANGE SIGNIFICANCE: As part of the significant overhaul of Section R310, Sections R310.3 and R310.4 in the 2021 IRC consolidate and reorganize provisions for doors, window wells, area wells and related items, such as covers, ladders and steps. Section R310.3 now only covers doors that are used for emergency escape and rescue openings. Doors were specifically added to Section R310 in the 2015 edition of the code to reassure users that doors, which typically far exceed the required opening dimensions, are permitted for emergency escape and rescue. The door must be a side-hinged or sliding door.

Section R310.4 covers area wells, merging all the previous provisions for window wells and area wells. It is a general scoping paragraph that clarifies that the bottom of an emergency escape and rescue opening lower than the adjacent grade triggers the requirements for an area well. For below-grade emergency escape and rescue openings, an area well is required whether the opening is a door or a window. The term "window well" is no longer used in the provisions for emergency escape and rescue openings.

The significant technical change to the area well provisions serving either a door or a window is the addition of dimension requirements for steps. When the area well is more than 44 inches deep, the code requires either a ladder or steps for the occupant to safely climb from the area well in an emergency situation. Language confirming that the ladder or steps can encroach up to 6 inches into the required area well dimensions has been retained as an exception. The code still emphasizes that ladder or steps must be usable with the door or window in the open position. The code has always been very specific about ladder dimensions but silent on the step dimensions other than to say that they were not required to comply with the stair provisions in Section R311.7. The step dimensions for width and riser now parallel those for ladders and the tread depth is the minimum width for alternating tread devices and ships ladders in Sections R311.7.11 and R311.7.12. The minimum inside width is 12 inches with a minimum tread depth of 5 inches and a maximum riser height of 18 inches. There are many manufactured area (window) well units of various materials on the market with steps built in that meet these dimension requirements.

Changes to the area well requirements include:

- Section R310.3: The last sentence about area wells is deleted as redundant since the criteria for area wells is specifically addressed later in Section 310.4.
- Sections R310.3.2, R310.3.2.1 and R310.3.2.2: The separate area well requirements for doors are deleted. Window wells and area wells are merged into the area well provisions of Section R310.4.
- Section R310.4.1: The new title and revisions to the text for area well dimensions provide consistent terminology.
- Section R310.4.2: This section states when a ladder or steps are required. The sentence about the window not obstructing the steps or ladder is a safety feature that has been retained.
- Sections R310.4.2.1 and R310.4.2.2: The dimensions for ladders and steps have been moved into separate sections. The step dimensions are new as discussed above.
- Section R310.4.3: Since the code always allows approved alternative methods, the last phrase for an alternative means of area well drainage has been deleted.
- Section R310.4.4: The provisions for bars, grilles, covers and screens over area wells have been revised to coordinate with other sections. The term "special knowledge" has been removed from the operational constraints because it allows for too broad an interpretation. (Note that the term has been retained in Section R310.2.1 for operation of emergency escape and rescue openings.)

R310.5, R310.6, R310.7
Emergency Escape and Rescue Openings in Existing Buildings

CHANGE TYPE: Modification

CHANGE SUMMARY: Opening dimensions have been reduced for emergency escape and rescue openings for a basement remodel, basement addition and for a change of occupancy.

2021 CODE: ~~R310.2.5~~ <u>R310.5 Replacement windows **for emergency escape and rescue openings.**</u> Replacement windows installed in buildings meeting the scope of this code shall be exempt from ~~the maximum~~

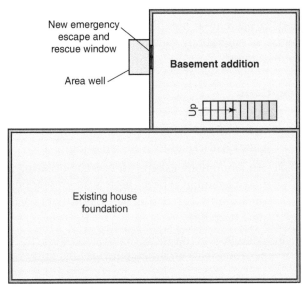

Basement addition with required emergency escape and rescue opening

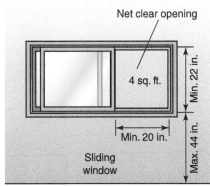

Reduced opening dimensions for emergency escape and rescue openings serving a basement addition or for a change of occupancy.

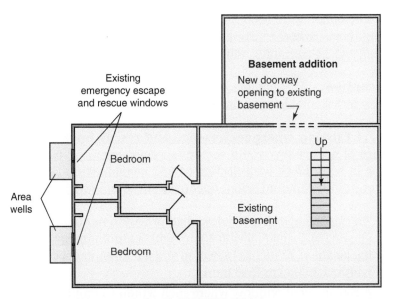

Basement addition with opening to existing basement

Access to an emergency escape and rescue opening is required for basement additions.

~~sill height requirements of Section R310.2.2 and the requirements of Section R310.2.1~~ <u>Sections R310.2 and R310.4.4</u>, provided that the replacement window meets the following conditions:

1. The replacement window is the manufacturer's largest standard size window that will fit within the existing frame or existing rough opening. The replacement window is of the same operating style as the existing window or a style that provides for an equal or greater window opening area than the existing window.
2. The replacement window is not part of a change of occupancy.

~~R310.5~~ <u>R310.6</u> Dwelling additions. Where dwelling additions contain sleeping rooms, an emergency escape and rescue opening shall be provided in each new sleeping room. Where dwelling additions have basements, an emergency escape and rescue opening shall be provided in the new basement.

Exceptions:
1. An emergency escape and rescue opening is not required in a new basement that contains a sleeping room with an emergency escape and rescue opening.
2. An emergency escape and rescue opening is not required in a new basement where there is an emergency escape and rescue opening in an existing basement that is accessed from the new basement.
3. <u>An operable window complying with Section 310.7.1 shall be acceptable as an emergency escape and rescue opening.</u>

~~R310.6~~ <u>R310.7</u> Alterations or repairs of existing basements. ~~An emergency escape and rescue opening is not required where existing basements undergo alterations or repairs.~~ **Exception:** New sleeping rooms created in an existing basement shall be provided with emergency escape and rescue openings in accordance with Section R310.1. <u>Other than new sleeping rooms, where existing basements undergo alterations or repairs an emergency escape and rescue opening is not required.</u>

Exception: <u>An operable window complying with Section 310.7.1 shall be acceptable as an emergency escape and rescue opening.</u>

<u>R310.7.1 Existing emergency escape and rescue openings.</u> <u>Where a change of occupancy would require an emergency escape and rescue opening in accordance with Section 310.1, operable windows serving as the emergency escape and rescue opening shall comply with the following:</u>

1. <u>An existing operable window shall provide a minimum net clear opening of 4 square feet (0.38 m^2) with a minimum net clear opening height of 22 inches (559 mm) and a minimum net clear opening width of 20 inches (508 mm).</u>
2. <u>A replacement window where such window complies with both of the following:</u>
2.1 <u>The replacement window meets the size requirements in Item 1.</u>

2.2 The replacement window is the manufacturer's largest standard size window that will fit within the existing frame or existing rough opening. The replacement window shall be permitted to be of the same operating style as the existing window or a style that provides for an equal or greater window opening area than the existing window.

CHANGE SIGNIFICANCE: The provisions for replacing existing windows in locations requiring an emergency escape and rescue opening have been relocated from Section R310.2.5 to their own Section R310.5. Changes to the text are editorial. The section title has changed to clarify that it is about emergency escape and rescue openings. The exemptions and criteria for replacement windows remain the same, but information has been shifted as part of the reorganization. When replacing windows in an existing dwelling unit, such as in a bedroom where an emergency escape and rescue opening is required, the intent is that the replacement window is no less safe than the existing window. The replacement must be the manufacturer's largest standard size window that will fit within the existing opening and be of the same operating style as the existing window or equivalent.

Section R310.6 deals with requirements for dwelling additions. The previous language requiring an emergency escape and rescue opening for new bedrooms and for a new basement that is part of the addition has been retained. A new Exception 3 references new Section R310.7.1 providing that the emergency escape and rescue opening can be reduced in size to a clear area not less than 4 square feet with a minimum height of 22 inches and minimum width of 20 inches.

Section R310.7 regarding alterations or repairs of existing basements has been restructured to clarify the application. Only if new sleeping rooms are added to an existing basement does an emergency escape and rescue opening need to be installed. Previously, this was written as an exception, but is more appropriately placed in the general rule. Otherwise, remodeling of an existing basement does not retroactively trigger the installation of an emergency escape opening as is required for new basements. A new exception addresses a situation where a new sleeping room is being created in an existing basement and there is an existing window in that new bedroom area. The code now allows some compromise on the opening size of this existing window by referencing a new provision in Section R310.7.1. In this case, the net clear opening area needs to be at least 4 square feet (rather than 5 square feet for a new grade floor opening) with a minimum height of 22 inches (rather than 24 inches for a new installation) and a minimum width of 20 inches.

Section R310.7.1 deals with existing windows where there is a change of occupancy and an emergency escape and rescue opening is required. In this case, the builder can take advantage of the reduced opening size of 4 square feet for the existing opening. While this section is specific to changes of occupancy, it is also referenced for new additions and for alterations to existing basements as discussed previously. A change of use or occupancy is unusual in the IRC unless the building is changed to a use governed by the IBC, in which case R310.7.1 would not apply – the building would need to comply with the IBC under changes of use or occupancy. A change of use under the IRC typically involves changing from a townhouse unit to a live/work unit or changing a single-family dwelling to a bed and breakfast operation. The new provisions are based on language in the *International Existing Building Code* (IEBC).

R311.7, R311.8
Stairways and Ramps

CHANGE TYPE: Clarification

CHANGE SUMMARY: The provisions of Sections R311.7 and R311.8 apply only to stairways and ramps within or serving a building, porch or deck.

2021 CODE: R202 STAIRWAY. One or more flights of stairs, either interior or exterior, with the necessary landings and connecting platforms to form a continuous and uninterrupted passage from one level to another ~~within or attached to a building, porch or deck~~.

R311.7 Stairways. <u>Where required by this code or provided, stairways shall comply with this section.</u>

<u>**Exceptions:**</u>

1. <u>Stairways not within or attached to a building, porch or deck</u>
2. <u>Stairways leading to non-habitable attics</u>
3. <u>Stairways leading to crawl spaces.</u>

Stairways attached to porches and decks must meet the requirements of Section R311.7.

R311.8 Ramps. Where required by this code or provided, ramps shall comply with this section.

> **Exception:** Ramps not within or attached to a building, porch or deck.

CHANGE SIGNIFICANCE: Scoping language has been added to Section R311.7 Stairways and R311.8 Ramps to clarify that the associated elements apply when they are within or attached to a building, porch or deck. Previously, similar language was found in Section R202 in the definition of a stairway as being "within or attached to a building, porch or deck." The definition has been changed to delete the scoping language and place it in Section R311. The definition now matches the IBC definition for consistency. The new text in Section R311 clarifies the intent and matches what has been common practice in most parts of the country. Stair safety is equally important to occupants of the dwelling whether it is an interior or exterior stair associated with the dwelling or if it is a stair that serves a deck or porch on any side of the house. Some voiced concern that because the stair provisions were located in the means of egress section of the code, that only those stairs leading to or from the one required exterior egress door were being regulated. That has never been the intent and the scoping is now clear.

There are now three exceptions to the stairway provisions in Section R311.7. The first, as discussed above, states that stairways that are not within or attached to a building, porch or deck are not regulated by this section. Exceptions 2 and 3 exempt stairs leading to nonhabitable attics and crawl spaces. This now specifically allows the common practice of providing a ladder to crawl spaces and a drop-down ladder/stair device for accessing a nonhabitable attic area.

R311.7.7
Stairway and Landing Walking Surface

CHANGE TYPE: Modification

CHANGE SUMMARY: A new exception allows steeper slopes for exterior landings that also serve to drain surface water away from the building.

2021 CODE: R311.7.7 Stairway walking surface. The walking surface of treads and landings of stairways shall be sloped not steeper than one unit vertical in 48 ~~inches~~ <u>units</u> horizontal (2-percent slope).

> <u>**Exception:** Where the surface of a landing is required elsewhere in the code to drain surface water, the walking surface of the landing shall be sloped not steeper than 1 unit vertical in 20 units horizontal (5-percent slope) in the direction of travel.</u>

CHANGE SIGNIFICANCE: The code regulates the slope of the ground (grade) and impervious surfaces, such as concrete patios, to provide minimum drainage requirements to direct surface water away from the structure. Section R401.3 generally requires a minimum slope for finish grade of 6-inch fall in the first 10 feet away from the building. This is a minimum 5 percent average slope. Impervious surfaces, such as concrete or brick paving, are more efficient at draining water and the code reduces the minimum slope accordingly to 2 percent or 2 inches fall in 10 feet. Sections R311.3 and R311.7.7 set maximum slopes for interior and exterior landings and stairs to provide comfortable and safe walking surfaces, and still provide drainage for exterior surfaces. The new exception addresses a concern that there are situations where the minimum slope for drainage might conflict with the maximum slope allowed for an exterior landing. In other words, the installation would require a steeper slope than is permitted for a landing when the landing is part of the drainage platform. Another possible installation difficulty would be to achieve a minimum slope of 2 percent for drainage while matching that with a maximum slope of 2 percent for walking, precision that may present some challenges. To satisfy the surface water drainage requirements under these circumstances, an exterior landing will now be permitted to slope as much as the minimum required for finish grade (ground) of 5 percent slope. This 5 percent also matches the maximum slope for an accessible route in the accessibility provisions of Chapter 11 of the IBC.

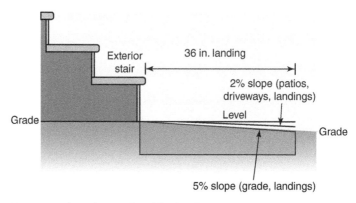

5 percent slope is permitted for landings that are required to provide drainage away from the dwelling.

R312.2 Window Fall Protection

CHANGE TYPE: Clarification

CHANGE SUMMARY: The revised language clarifies that measurements for determining the need for fall protection are taken to the bottom of the clear opening of the window.

2021 CODE: R312.2 Window fall protection. Window fall protection shall be provided in accordance with Sections R312.2.1 and R312.2.2.

R312.2.1 Window sills opening height. In dwelling units, where the top bottom of the sill clear opening of an operable window opening is located less than 24 inches (610 mm) above the finished floor and greater than 72 inches (1829 mm) above the finished grade or other surface below on the exterior of the building, the operable window shall comply with one of the following:

1. Operable window openings will not allow a 4-inch-diameter (102 mm) sphere to pass through where the openings are in their largest opened position.
2. Operable windows are provided with window opening control devices or fall prevention devices that comply with ASTM F2090.
3. Operable windows are provided with window opening control devices that comply with Section R312.2.2.

R312.2.2 Window opening control devices. Emergency escape and rescue openings. Window opening control devices shall comply with ASTM F2090. The Where an operable window serves as an emergency escape and rescue opening, a window opening control device or fall prevention device, after operation to release the control device or fall prevention device allowing the window to fully open, shall not reduce the net clear opening area of the window unit to less than the area required by Sections R310.2.1 and R310.2.2.

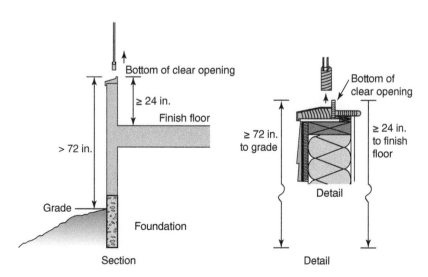

Measurements for window fall protection are taken to the bottom of the clear opening.

CHANGE SIGNIFICANCE: The provisions for window fall protection intend to reduce the number of injuries to children resulting from falls through windows. These provisions first appeared in the 2006 edition of the IRC. Since then, the height of the opening above grade deemed to pose a hazard has been greater than 72 inches. Where the height has exceeded 72 inches, the minimum height of the opening above the finished floor has been set at 24 inches, unless other protection, such as an opening control device, was provided. The wording, particularly how the measurements for this section are taken, have changed from time to time. In the original text, the section was titled "Window Sills" and the measurements were taken to the opening. In the 2015 edition, the language changed so the measurement was taken to the window sill. A sill, if it exists, is not necessarily the lowest point of the window opening. The reference to window sill in the title and the text has been removed and measurements are taken to the bottom of the clear opening.

The other change to this section relates to window opening control devices and fall prevention devices that have been alternates to the opening height requirements for child fall prevention. When used, these devices must be in compliance with ASTM F2090, *Standard Specification for Window Fall Prevention Devices with Emergency Escape (Egress) Release Mechanisms*. As the title indicates, these devices are approved for use on emergency escape and rescue windows when conforming to this standard. Changes to the code text are meant to clarify the use of these devices and emphasize that they are acceptable for use on emergency escape and rescue windows. Text for opening control devices and fall prevention devices being allowed for emergency escape and rescue windows also appears in Section R310.

R314.3 Smoke Alarm Locations

CHANGE TYPE: Modification

CHANGE SUMMARY: A new location requirement for smoke alarms addresses high ceilings adjacent to hallways serving bedrooms.

2021 CODE: R314.3 Location. Smoke alarms shall be installed in the following locations:

1. In each sleeping room.
2. Outside each separate sleeping area in the immediate vicinity of the bedrooms.
3. On each additional story of the dwelling, including basements and habitable attics and not including crawl spaces and uninhabitable attics. In dwellings or dwelling units with split levels and without an intervening door between the adjacent levels, a smoke alarm installed on the upper level shall suffice for the adjacent lower level provided that the lower level is less than one full story below the upper level.
4. ~~Smoke alarms shall be installed not~~ <u>Not</u> less than 3 feet (914 mm) horizontally from the door or opening of a bathroom that contains a bathtub or shower unless this would prevent placement of a smoke alarm required by this section.
5. <u>In the hallway and in the room open to the hallway in dwelling units where the ceiling height of a room open to a hallway serving bedrooms exceeds that of the hallway by 24 inches (610 mm) or more.</u>

R314.3.1 Installation near cooking appliances. Smoke alarms shall not be installed in the following locations unless this would prevent placement of a smoke alarm in a location required by Section R314.3.

Hallway serving bedrooms and adjacent higher ceiling both require smoke alarms.

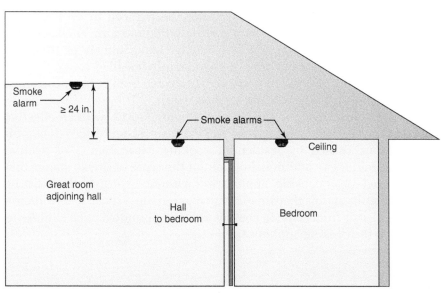

Section drawing

Smoke alarm locations.

1. Ionization smoke alarms shall not be installed less than 20 feet (6096 mm) horizontally from a permanently installed cooking appliance.
2. Ionization smoke alarms with an alarm-silencing switch shall not be installed less than 10 feet (3048 mm) horizontally from a permanently installed cooking appliance.
3. Photoelectric smoke alarms shall not be installed less than 6 feet (1828 mm) horizontally from a permanently installed cooking appliance.
4. <u>Smoke alarms listed and marked "helps reduce cooking nuisance alarms" shall not be installed less than 6 feet (1828 mm) horizontally from a permanently installed cooking appliance.</u>

CHANGE SIGNIFICANCE: Since its beginning, the IRC has required smoke alarms to be located in each bedroom, in the area adjacent to bedrooms (often a hallway), and on each level including basements. As an alternative to each level, where an open split-level design occurs, the code permits the smoke alarm on the upper level to serve both levels. This recognizes that smoke rises and the alarm on the upper level will react quickly to smoke that originates on the level below. A similar situation has prompted a change to the location requirements in the 2021 IRC. If the hallway serving the bedrooms opens to a great room with a high ceiling, the required smoke alarm in the hall may not sound immediately as smoke rises and accumulates at the upper ceiling of the great room. Previously, the code did not require a smoke alarm at the higher level. If the upper ceiling in this case is greater than 24 inches above the hall ceiling, the code now requires an additional smoke alarm on the upper ceiling in addition to the one in the hall. This will provide earlier notification for occupants to safely exit the building.

The 2015 edition of the IRC introduced measures to reduce nuisance alarms by requiring horizontal separation of smoke alarms from cooking appliances and bathrooms. The minimum separation requirements from permanently installed cooking appliances vary based on the type of smoke alarm installed. Ionization smoke alarms generally require a separation distance of 20 feet, but that distance may be reduced to 10 feet if the smoke alarm has an alarm-silencing switch. Photoelectric smoke alarms are less susceptible to activation by cooking vapors and are permitted to be located as close as 6 feet from a permanently installed cooking appliance. The intent is to regulate separation distance from built-in cook tops and ovens as well as stand-alone kitchen ranges.

Smoke alarms listed to the new edition of UL 217 (with an effective date of May 29, 2020) are required to pass tests designed to reduce nuisance alarms caused by residential cooking. New item 4 of Section R314.3.1 provides an additional option for the types of smoke alarms that can be used near cooking appliances, without changing the existing options. Smoke alarms that are identified as having resistance to common nuisance alarms from cooking sources are now permitted to be as close as 6 feet from the cooking appliance.

CHANGE TYPE: Modification

CHANGE SUMMARY: Repairs to an existing fuel-fired mechanical system now trigger the retroactive requirements for carbon monoxide alarms.

2021 CODE: R315.2.2 Alterations, repairs and additions. Where alterations, repairs or additions requiring a permit occur, the individual dwelling unit shall be equipped with carbon monoxide alarms located as required for new dwellings.

Exceptions:

1. Work involving the exterior surfaces of dwellings, such as the replacement of roofing or siding, or the addition or replacement of windows or doors, or the addition of a porch or deck.
2. Installation, alteration or repairs of plumbing ~~or mechanical~~ systems.
3. <u>Installation, alteration or repairs of mechanical systems that are not fuel fired.</u>

CHANGE SIGNIFICANCE: Requirements for carbon monoxide alarms were introduced in the 2009 edition of the IRC to reduce accidental injury and deaths from carbon monoxide poisoning. Because the source of unsafe levels of carbon monoxide in the home is typically from faulty operation of a fuel-fired furnace or water heater, or from the exhaust of an automobile, the provisions only apply to dwellings containing fuel-fired appliances or having an attached garage that communicates with

R315.2.2
Carbon Monoxide Alarms

Carbon monoxide alarm.

Repair of a fuel-fired furnace.

the residence. Carbon monoxide accumulates in the body over time relative to its concentration in the air. Accordingly, carbon monoxide alarms sound an alarm based on the concentration of carbon monoxide and the amount of time that certain levels are detected, simulating an accumulation of the toxic gas in the body. High levels of carbon monoxide will trigger an alarm within a short time, while lower levels must be present over a longer time period for the alarm to sound. This design prevents false-positive alarms.

Because carbon monoxide poisoning deaths often occur when the occupants are sleeping, the IRC requires carbon monoxide alarms to be located in the areas outside of and adjacent to bedrooms. Similar to the smoke alarm provisions, carbon monoxide alarms are not only required in new dwellings but are also required in existing dwelling units when triggered by construction work on the existing dwelling, one of only two provisions in the code that are retroactive. When work requiring a permit occurs, alarms must be installed in the locations prescribed by the code. Matching the smoke alarm provisions, exterior work requiring a permit, such as roofing, siding, windows, doors, porches and decks, does not trigger the retroactive carbon monoxide alarm requirements.

The second exception to the retroactive provisions has exempted the installation, alteration or repairs of plumbing or mechanical systems. New to the 2021 IRC, alteration or repairs to mechanical systems are only exempt if they do not include fuel-fired equipment. Mechanical systems often include forced air units and water heaters that are fuel burning and therefore subject to carbon monoxide leakage in the event of faulty operation. Consensus was that fuel-fired equipment that is being altered, repaired or replaced and requires a permit should be subject to the carbon monoxide alarm provisions. The language for this exception has been changed to only apply to mechanical systems that are not fuel fired.

Significant Changes to the IRC 2021 Edition   R317.1 ■ Protection of Wood Against Decay   **65**

R317.1
Protection of Wood Against Decay

CHANGE TYPE: Clarification

CHANGE SUMMARY: The provisions of Section R317.1 have been revised and reorganized for clarification.

2021 CODE: R317.1 Location required. Protection of wood and wood-based products from decay shall be provided in the following locations by the use of naturally durable wood or wood that is preservative-treated in accordance with AWPA U1.

1. ~~Wood joists or the bottom of a wood structural floor where closer than 18 inches (457 mm) or, wood girders where closer than 12 inches (305 mm) to the exposed ground , in~~ In crawl spaces or unexcavated area located within the periphery of the building foundation~~.~~, wood joists or the bottom of a wood structural floor where closer than 18 inches (457 mm) to exposed ground, wood girders where closer than 12 inches (305 mm) to exposed ground, and wood columns where closer than 8 inches (204 mm) to exposed ground.
2. Wood framing members, including columns, that rest directly on concrete or masonry exterior foundation walls and are less than 8 inches (203 mm) from the exposed ground.

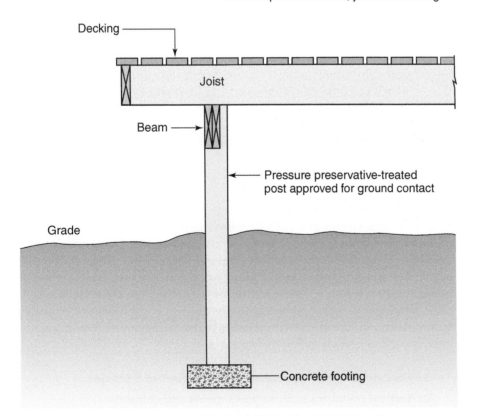

Protection against decay for wood structural members exposed to the weather.

3. Sills and sleepers on a concrete or masonry slab that is in direct contact with the ground unless separated from such slab by an impervious moisture barrier.

4. The ends of wood girders entering exterior masonry or concrete walls having clearances of less than ½ inch (12.7 mm) on tops, sides and ends.

5. Wood siding, sheathing and wall framing on the exterior of a building having a clearance of less than 6 inches (152 mm) from the ground or less than 2 inches (51 mm) measured vertically from concrete steps, porch slabs, patio slabs and similar horizontal surfaces exposed to the weather.

6. Wood structural members supporting moisture-permeable floors or roofs that are exposed to the weather, such as concrete or masonry slabs, unless separated from such floors or roofs by an impervious moisture barrier.

7. Wood furring strips or other wood framing members attached directly to the interior of exterior masonry walls or concrete walls below grade except where an approved vapor retarder is applied between the wall and the furring strips or framing members.

8. <u>Portions of wood structural members that form the structural supports of buildings, balconies, porches or similar permanent building appurtenances where those members are exposed to the weather without adequate protection from a roof, eave, overhang or other covering that would prevent moisture or water accumulation on the surface or at joints between members.</u>

   <u>**Exception:** Sawn lumber used in buildings located in a geographical region where experience has demonstrated that climatic conditions preclude the need to use naturally durable or preservative-treated wood where the structure is exposed to the weather.</u>

9. <u>Wood columns in contact with basement floor slabs unless supported by concrete piers or metal pedestals projecting at least 1 inch (25 mm) above the concrete floor and separated from the concrete pier by an impervious moisture barrier.</u>

R317.1.1 Field treatment. *[No change]*

R317.1.2 Ground contact. *[No change]*

R317.1.3 ~~**Geographical areas.**~~ ~~In geographical areas where experience has demonstrated a specific need, approved naturally durable or pressure-preservative-treated wood shall be used for those portions of wood members that form the structural supports of buildings, balconies, porches or similar permanent building appurtenances where those members are exposed to the weather without adequate protection from a roof, eave, overhang or other covering that would prevent moisture or water accumulation on the surface or at joints between members. Depending on local experience, such members typically include:~~

   1. ~~Horizontal members such as girders, joists and decking.~~
   2. ~~Vertical members such as posts, poles and columns.~~
   3. ~~Both horizontal and vertical members.~~

R317.1.4 Wood columns. ~~Wood columns shall be approved wood of natural decay resistance or approved pressure-preservative-treated wood.~~

~~**Exceptions:**~~

1. ~~Columns exposed to the weather or in basements where supported by concrete piers or metal pedestals projecting 1 inch (25 mm) above a concrete floor or 6 inches (152 mm) above exposed earth and the earth is covered by an approved impervious moisture barrier.~~

2. ~~Columns in enclosed crawl spaces or unexcavated areas located within the periphery of the building where supported by a concrete pier or metal pedestal at a height more than 8 inches (203 mm) from exposed earth and the earth is covered by an impervious moisture barrier.~~

3. ~~Deck posts supported by concrete piers or metal pedestals projecting not less than 1 inch (25 mm) above a concrete floor or 6 inches (152 mm) above exposed earth.~~

R317.1.5 Exposed glued-laminated timbers. ~~The portions of glued-laminated timbers that form the structural supports of a building or other structure and are exposed to weather and not properly protected by a roof, eave or similar covering shall be pressure treated with preservative, or be manufactured from naturally durable or preservative-treated wood.~~

CHANGE SIGNIFICANCE: The provisions for protection of wood against decay have not changed significantly for many code editions. This update accomplishes a much-needed overhaul. There is no intent to significantly change the technical content of Section R317, only to clarify the requirements, clean up some errors in syntax, and correct some misleading text. The revised text more accurately reflects the way this section has been applied.

The following is a summary of the changes:

- Section R317.1 Item 1: The wording is rearranged for readability. Similar to floor framing and girders, columns are given a required clearance from exposed earth in crawl spaces, which is generally consistent with former Exception 2 to Section R317.1.4. The moisture barrier language related to columns in crawl spaces has been removed because that should be governed by the crawl space provisions in Section R408.
- Section R317.1 Item 2: This item now includes columns with other wood framing members since the columns Section R317.1.4 has been deleted.
- Section R317.1 Item 8: The language comes from the deleted Section R317.1.3 (Geographical areas), which seemed to require wood members exposed to the weather to be treated or naturally durable only if a need was demonstrated. The code now requires protection as the general rule and provides an exception for arid geographical regions where it has been demonstrated that the protection is not needed. Relocating this to the list of locations in R317.1 and providing an

exception clarifies the requirement. The new exception also makes clear that it only applies to sawn lumber, not engineered wood products, which should be protected when exposed to the weather regardless of climate or geographic location, and manufacturers' recommendations require protection.

- Section R317.1 Item 9: This new item is necessary to preserve the reduced clearance for columns above basement floor slabs consistent with part of deleted Exception 1 to Section R317.1.4. It provides for as little as 1 inch of clearance if on a metal pedestal and 1 inch of clearance on a concrete pier if it is separated from the pier by an impervious moisture barrier, since concrete is porous and will allow wicking of moisture more readily.
- Section R317.1.3: The section has been deleted, and the text revised and moved to the new Item 8 of Section R317.1. The descriptors in Items 1 through 3 about horizontal and vertical members have been removed because they were considered to be commentary rather than code language.
- Section R317.1.4: This section has been deleted because it was unnecessarily confusing and contained errors in syntax, making it difficult to apply. The charging language seemed to require all columns, regardless of location (including interior locations), to be treated. That is not the intent of the code.
  - Exception 1: This deleted exception seemed to exempt all columns exposed to the weather, which was not the intent. The rest of Exception 1 had criteria which conflicted with the current IBC and also seemed to conflict with Exception 2.
  - Exception 2: This has been clarified and moved to revised Item 1 of Section R317.1.
  - Exception 3: This exception has been deleted because it seemed to exempt any deck posts that are supported by piers or pedestals extending 1 inch above concrete or 6 inches above exposed earth. Best practices dictate that any deck post exposed to the weather should be protected against decay regardless of clearance to a slab or ground.
- Section R317.1.5: This section has been deleted because glued laminated timber is covered under the scope of new Item 8 of Section R317.1 and the current language is therefore redundant.

R320 Accessibility

CHANGE TYPE: Modification

CHANGE SUMMARY: The accessibility provisions for live/work units and owner-occupied lodging houses constructed under the IRC are clarified.

2021 CODE: R202 DEFINITIONS

<u>**LIVE/WORK UNIT.** A dwelling unit or sleeping unit in which a significant portion of the space includes a nonresidential use that is operated by the tenant.</u>

<u>**SLEEPING UNIT.** A single unit that provides rooms or spaces for one or more persons, includes permanent provisions for sleeping and can include provisions for living, eating and either sanitation or kitchen facilities but not both. Such rooms and spaces that are also part of a dwelling unit are not sleeping units.</u>

**SECTION R320
ACCESSIBILITY**

R320.1 Scope. Where there are four or more dwelling units or sleeping units in a single structure, the provisions of Chapter 11 of the *International Building Code* for Group R-3 shall apply.

<u>**Exception:** Owner-occupied lodging houses with five or fewer guestrooms are not required to be accessible.</u>

Live/work units

Live/work units must comply with the accessibility requirements of the referenced sections in the IBC.

R320.1.1 Guestrooms. ~~A dwelling with guestrooms shall comply with the provisions of Chapter 11 of the International Building Code for Group R-3. For the purpose of applying the requirements of Chapter 11 of the *International Building Code,* guestrooms shall be considered to be sleeping units.~~

> **Exception:** ~~Owner-occupied lodging houses with five or fewer guestrooms constructed in accordance with the International Residential Code are not required to be accessible.~~

R320.2 Live/work units. In live/work units, the nonresidential portion shall be accessible in accordance with Sections 508.5.9 and 508.5.11 of the *International Building Code.* In a structure where there are four or more live/work units, the dwelling portion of the live/work unit shall comply with Section 1108.6.2.1 of the *International Building Code.*

CHANGE SIGNIFICANCE: The design and construction of facilities for accessibility for individuals with disabilities is covered in Chapter 11 of the IBC. In general, IRC buildings do not require accessibility and accessible features are optional. Section R320 provides that where there are four or more dwelling units or sleeping units in a single structure, the accessibility provisions of IBC Chapter 11 apply. This is applicable to townhouses with four or more townhouse units. However, exceptions in IBC Chapter 11 exempt multilevel dwelling units without elevators, meaning that in most cases townhouses are exempt from the accessibility requirements.

The new language in Section R320 intends to clarify the accessibility requirements as they apply to live/work units in townhouses and owner-occupied lodging houses with no more than five sleeping units (typically bed and breakfast establishments), both allowed to be constructed under the exceptions to the scope of the IRC. The existing language in the IRC has not only exempted these lodging houses from accessibility requirements, but also included some confusing text requiring accessibility for dwellings with guestrooms followed by an exception for those IRC lodging houses. The confusing language has been removed and the exemption for owner-occupied lodging houses with five or fewer guestrooms appears as an exception to the general rule on accessibility.

The second change to this section addresses accessibility for live/work units to correspond to the referenced provisions in Sections 508 and Chapter 11 of the IBC. The exception to the scope in the IRC has stated that live/work units in townhouses must comply with the live/work unit provisions in Section 508 in the IBC. The new language in the IRC intends to strengthen and emphasize that link. For example, a definition for live/work unit has been added to the IRC to correlate with the IBC definition. In new Section R320.2, the code repeats the IBC language that the nonresidential portion of the live/work unit shall be accessible. In addition, where there are four or more live/work units, the dwelling portion of the live/work unit needs to comply with Section 1108.6.2.1 of the IBC. While this would seem to imply mandatory rules related to accessibility (Type B units), the IBC again exempts multilevel dwelling units without elevators.

R323 Storm Shelters

CHANGE TYPE: Addition

CHANGE SUMMARY: Added guidance on the design of storm shelters is placed in Section R323.

2021 CODE: R202 DEFINITIONS

<u>**STORM SHELTER.** A building, structure or portion thereof, constructed in accordance with ICC 500 and designated for use during a severe windstorm event, such as a hurricane or tornado.</u>

<u>**R106.1.5 Information on storm shelters.** Construction documents for storm shelters shall include the information required in ICC 500.</u>

**SECTION R323
STORM SHELTERS**

R323.1 General. This section applies to storm shelters where constructed as separate detached buildings or where constructed as safe rooms within buildings for the purpose of providing refuge from storms that produce high winds, such as tornados and hurricanes. In addition to other applicable requirements in this code, storm shelters shall be constructed in accordance with ICC ~~/NSSA~~-500.

Exterior of storm shelter.

ICC 500.

Single-family storm shelter.

R323.1.1 Sealed documentation. The construction documents for all structural components and impact protective systems of the storm shelter shall be prepared and sealed by a registered design professional indicating that the design meets the criteria of ICC-500.

> **Exception:** Storm shelters, structural components and impact-protective systems that are listed and labeled to indicate compliance with ICC-500.

CHANGE SIGNIFICANCE: Impact-protective systems of structures intended as residential storm shelters have failed prematurely when they do not meet the testing requirements of ICC 500 *Standard for the Design and Construction of Storm Shelters*. In some cases, the structures have been placed above ground where they were not designed for loads created by tornadoes or hurricanes. Reports of failures associated with residential storm shelters that are not designed and constructed in accordance with ICC 500 underscore the importance of these new provisions.

Failures have not occurred in residential shelters engineered and certified as residential storm shelters. The provisions of ICC 500 cannot be met by prescriptive methods in the IRC and require the expertise of a registered design professional.

Section R323.1 requires that storm shelters comply with ICC 500. By adding a definition of a storm shelter to the IRC, adding a requirement in Section R106, Construction documents, for details required by ICC 500 and a requirement for sealed plans providing structural and impact protection system design, these shelters should withstand tornadoes or hurricanes keeping deaths and injuries to a minimum. The exception in Section R323.1.1 allows listed and labelled shelter designs to be submitted without an individual design as these shelters have a third-party review process checking the design's resistance to high wind loads and impacts.

R324.3 Photovoltaic Systems

CHANGE TYPE: Modification

CHANGE SUMMARY: Building-integrated photovoltaic (BIPV) systems meeting the specified criteria do not require firefighter access pathways and setbacks.

2021 CODE: R324.3 Photovoltaic systems. Photovoltaic systems shall be designed and installed in accordance with Sections R324.3.1 through R324.7.1, ~~NFPA 70~~ and the manufacturer's installation instructions. <u>The electrical portion of solar PV systems shall be designed and installed in accordance with NFPA 70.</u>

R324.3.1 Equipment listings. Photovoltaic panels and modules shall be listed and labeled in accordance with UL 1703 <u>or with both UL 61730-1 and UL 61730-2</u>. Inverters shall be listed and labeled in accordance with UL 1741. Systems connected to the utility grid shall use inverters listed for utility interaction. <u>Mounting systems listed and labeled in accordance with UL 2703 shall be installed in accordance with the manufacturer's installation instructions and their listings.</u>

R324.5 Building-integrated photovoltaic systems. Building-integrated photovoltaic <u>(BIPV)</u> systems that serve as roof coverings shall be designed and installed in accordance with Section R905.

R324.5.1 Photovoltaic shingles. Photovoltaic shingles shall comply with Section R905.16.

R324.5.2 Fire classification. Building-integrated photovoltaic systems shall have a fire classification in accordance with Section R902.3.

<u>**R324.5.3 BIPV roof panels.** BIPV roof panels shall comply with Section R905.17.</u>

Building-integrated photovoltaic system.

R324.6 Roof access and pathways. Roof access, pathways and setback requirements shall be provided in accordance with Sections R324.6.1 through R324.6.2.1. Access and minimum spacing shall be required to provide emergency access to the roof, to provide pathways to specific areas of the roof, provide for smoke ventilation opportunity areas, and to provide emergency egress from the roof.

Exceptions:

1. Detached, nonhabitable structures, including but not limited to detached garages, parking shade structures, carports, solar trellises and similar structures, shall not be required to provide roof access.
2. Roof access, pathways and setbacks need not be provided where the code official has determined that rooftop operations will not be employed.
3. These requirements shall not apply to roofs with slopes of two units vertical in 12 units horizontal (17-percent slope) or less.
4. <u>BIPV systems listed in accordance with Section 690.12(B) (2) of NFPA 70, where the removal or cutting away of portions of the BIPV system during firefighting operations have been determined to not expose a firefighter to electrical shock hazards.</u>

R324.6.2.2 <u>R324.6.3</u> Emergency escape and rescue openings. Panels and modules installed on dwellings shall not be placed on the portion of a roof that is below an emergency escape and rescue opening. A pathway not less than 36 inches (914 mm) wide shall be provided to the emergency escape and rescue opening.

Exception: <u>BIPV systems listed in accordance with Section 690.12(B)(2) of NFPA 70, where the removal or cutting away of portions of the BIPV system during firefighting operations have been determined to not expose a firefighter to electrical shock hazards.</u>

R905.16.4 Material standards. Photovoltaic shingles shall be listed and labeled in accordance with UL ~~1703~~ <u>7103 or with both UL 61730-1 and UL 61730- 2</u>.

R905.17.5 Material standards. BIPV roof panels shall be listed and labeled in accordance with UL ~~1703~~ <u>7103 or with both UL 61730-1 and UL 61730-2</u>.

CHANGE SIGNIFICANCE: Photovoltaic (PV) solar energy systems offer property owners the ability to generate their own electricity and, in many cases, sell excess electricity back to the utility provider. These PV systems have proliferated in recent years and are now commonly seen on rooftops of residential buildings in many areas of the country.

The roof access and pathway provisions for firefighters first appeared in the 2018 IRC. Access, pathways and ridge setbacks are provided so firefighters can perform manual ventilation by cutting one or more holes in a building roof and other rooftop activities as necessary. The provisions do not apply to detached garages and similar buildings or to dwellings with low-slope roofs of 2:12 or less. There is also a provision that exempts dwellings where the building official has determined that the fire department will not perform rooftop operations for that property.

The technology of solar roofs has been advancing with new materials and methods, particularly in the area of BIPV systems. Unlike conventional PV panel systems mounted above the roof surface, BIPV systems are integrated into the finished roof surface, and do not present significant trip hazards or physical obstacles to equipment. There are BIPV systems available today that have been shown through testing to not present electrical hazards to firefighters even when cutting into them during ventilation operations. The code now recognizes these safety features in two new exceptions to the firefighter access and occupant emergency escape rooftop pathways. Where listed to the recognized standard that determines that the BIPV materials do not expose a firefighter to electrical shock hazards including during cutting operations, the requirement to provide for access and pathways has been deleted. Similarly, a clear pathway is no longer required below an emergency escape and rescue opening where these BIPV systems are installed. The BIPV is the roof surface that provides the walking surface similar to other conventional roofing materials.

R326
Habitable Attics

CHANGE TYPE: Modification

CHANGE SUMMARY: The habitable attic provisions have been placed in new Section R326 and new restrictions limit their area and require an automatic fire sprinkler system.

2021 CODE: R325.1 General. Mezzanines shall comply with Sections R325 through R325.5. ~~Habitable attics shall comply with Section R325.6.~~

~~**R325.6 Habitable attic.** A habitable attic shall not be considered a story where complying with all of the following requirements:~~

~~1. The occupiable floor area is not less than 70 square feet (17 m²), in accordance with Section R304.~~

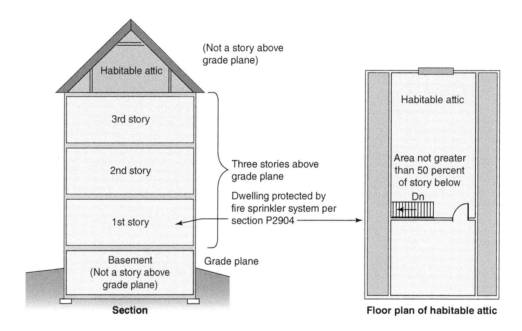

Section

Floor plan of habitable attic

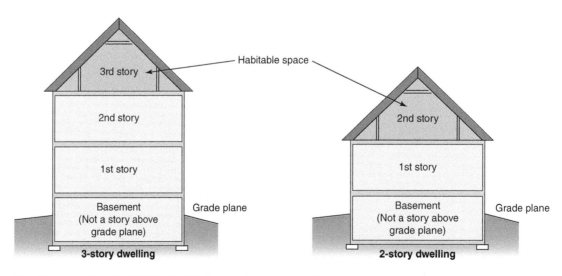

Dwellings meeting the IRC limit of 3 stories above grade plane.

~~2. The occupiable floor area has a ceiling height in accordance with Section R305.~~

~~3. The occupiable space is enclosed by the roof assembly above, knee walls (if applicable) on the sides and the floor-ceiling assembly below.~~

~~4. The floor of the occupiable space shall not extend beyond the exterior walls of the floor below.~~

SECTION R326
HABITABLE ATTICS

R326.1 General. Habitable attics shall comply with Sections R326.2 and R326.3.

R326.2 Minimum Dimensions. A habitable attic shall have a floor area in accordance with Section R304 and a ceiling height in accordance with Section R305.

R326.3 Story Above Grade Plane. A habitable attic shall be considered a story above grade plane.

> **Exception:** A habitable attic shall not be considered to be a story above grade plane provided that the habitable attic meets all of the following:
>
> 1. The aggregate area of the habitable attic is either of the following:
>   1.1 Not greater than one-third of the floor area of the story below.
>   1.2 Not greater than one-half of the floor area of the story below where the habitable attic is located within a dwelling unit equipped with a fire sprinkler system in accordance with Section P2904.
> 2. The occupiable space is enclosed by the roof assembly above, knee walls, if applicable, on the sides and the floor-ceiling assembly below.
> 3. The floor of the habitable attic does not extend beyond the exterior walls of the story below.
> 4. Where a habitable attic is located above a third story, the dwelling unit or townhouse unit shall be equipped with a fire sprinkler system in accordance with Section P2904.

R326.4 Means of Egress. The means of egress for habitable attics shall comply with the applicable provisions of Section R311.

CHANGE SIGNIFICANCE: The term "habitable attic" first appeared in the definitions of the 2009 edition of the IRC. Although finishing off habitable space in an attic was not unheard of, the origin of the term "habitable attic" is unclear and it may have been created for use in the IRC. By definition, an "attic" is unfinished space, and it is not considered to be occupiable or habitable. The only purpose for introducing the term "habitable attic" was to add another usable level to a dwelling constructed under the IRC in addition to the maximum height in stories prescribed by the code. Since its introduction, a habitable attic has not been considered a story and has been permitted in addition to the maximum three stories above grade plane as allowed by the scope of the code. Where a dwelling or townhouse had a basement that was not a story above grade plane, identifying the top level as a habitable attic in addition to the three stories above grade plane created five usable or habitable levels. This has been a design option to benefit the designer, builder and property owner in constructing taller buildings under the IRC. There is at least a perceived advantage to building under the IRC as opposed to the IBC. Perhaps the biggest issue comes down to the installation of a fire sprinkler system, which is required in both model codes. The sprinkler provisions of the IRC have been amended out of local ordinances in many parts of the United States. Where a builder perceived an advantage to building under the IRC provisions rather than those of the IBC, a habitable attic may have provided that flexibility.

In most parts of the country, three-story houses are unusual and those exceeding three stories are rarer still. However, in some urban areas of the country, space for new construction is limited and it is desirable to build a taller building (or add to the height of an existing building) on a smaller footprint.

In the code editions since the 2009 IRC, the rules for a habitable attic have remained consistent and changes have been mostly editorial. Technical requirements have been removed from the definition and placed in a section near the end of Chapter 3. The space has had to meet the minimum room area and ceiling height for habitable spaces and be enclosed by the roof assembly and floor/ceiling assembly of the attic. The code has also required habitable attics to have a smoke alarm and an emergency escape and rescue opening.

Concern was expressed that the added habitable space above the third story creates a fire- and life-safety hazard for occupants because of the height above fire department access and the maximum reach of standard 35-foot extension ladders that may be used in a fire department response. Discussion has also centered around the differences between the IRC and IBC. The IBC does not have provisions for a "habitable attic." As in the IRC, an "attic" in the IBC is unfinished and is not occupiable or habitable space. If the attic is converted to habitable space, it is no longer an attic but becomes habitable space on another story because it then meets the definition of story. The addition of habitable attic in the 2009 IRC was purposeful in that it was adding an option to the IRC to avoid being under the scope of the IBC. There was no intent for the provisions in the two codes to match.

New to the 2021 IRC, Section R326.3 states that a habitable attic is a story above grade plane. The exceptions retain the concept of a habitable attic being allowed above the third story and not being considered a story above grade plane with further restrictions. This option is available only if the dwelling unit is protected with a fire sprinkler system in accordance with Section P2904 or NFPA 13D, and the "aggregate" area of the habitable attic does not exceed one-half of the area of the story below. These additional requirements are thought to mitigate the life-safety concerns for occupants in these taller buildings.

The limitation on "aggregate" area for a habitable attic is borrowed from the mezzanine provisions (which also have very limited use under the IRC but, like habitable attics, may be used to build taller buildings since they are not considered a story). Unlike mezzanines, attics typically match the area of the story below. Presumably the intent, though not stated, is that this 50 percent limitation only applies to the habitable area of the attic and that "aggregate" means the combined area of habitable space on that level. For example, storage areas and areas that do not meet the minimum ceiling height requirements are not considered habitable space and would not be included in the 50 percent calculation. On the other hand, the area of the story below would include both habitable space and other spaces for calculation purposes. Exception 1.1 (one-third area limitation) has no application because it applies to a building without fire sprinklers and sprinklers are required in all cases where a habitable attic is above the third story and is not considered a story. This conflict occurs because there were multiple public comments to the initial code change proposal that were approved at the public comment hearings.

The new code language presents the option to call habitable space in the attic above a one-story or two-story house a "habitable attic" but there is no advantage to doing so as a design option. As in previous editions of the IRC, including the editions before the term habitable attic was introduced to the code, and as is done in the IBC, when an attic of a one- or two-story house is finished into habitable space, it is no longer an attic and becomes habitable space. Whether or not it is considered a story is no longer an issue.

PART 3
Building Construction
Chapters 4 through 10

- **Chapter 4** Foundations
- **Chapter 5** Floors
- **Chapter 6** Wall Construction
- **Chapter 7** Wall Covering
- **Chapter 8** Roof-Ceiling Construction
- **Chapter 9** Roof Assemblies
- **Chapter 10** Chimneys and Fireplaces
   No changes addressed

Chapters 4 through 10 address the prescriptive methods for building foundations, floor construction, wall construction, wall coverings, roof construction, roof assemblies, chimneys, and fireplaces. Concrete, masonry, and wood foundations; retaining walls; supporting soil properties; surface drainage; and foundation dampproofing and drainage are found in Chapter 4. Chapters 5, 6, and 8 contain the construction provisions for floors and decks, walls, and roofs, respectively, with most of the provisions addressing light-frame construction. Chapter 7 addresses interior finishes, such as drywall and plaster installations, and exterior wall coverings, including water-resistive barriers, flashings, siding, and veneer, to provide a durable weather-resistant exterior. Chapter 9 covers the various waterproof roof assemblies, including roofing underlayment, roof eave ice barrier, flashings, asphalt shingles, and other roof coverings. Site-built masonry fireplaces and chimneys as well as prefabricated fireplaces and chimneys, including their weather-tight roof terminations, are addressed in the provisions of Chapter 10. ■

TABLE R403.1(1)
Footings Below Light-Frame Construction

R406.2
Foundation Waterproofing

R408.8
Vapor Retarder in Crawlspaces

R506.2.3
Vapor Retarders Under Concrete Slabs

R507
Deck Loads

R507.3
Deck Footings

R507.4
Deck Posts

R507.5
Deck Beams

R507.6
Deck Joists

R507.7
Decking

R507.10
Exterior Guards

TABLE R602.3(1)
Fasteners – Roof and Wall

TABLE R602.3(1)
Fasteners – Roof Sheathing

TABLE R602.3(2)
Alternate Attachments

R602.9
Cripple Walls

R602.10.1.2
Location of Braced Wall Lines

R602.10.2.2
Location of Braced Wall Panels (BWPs)

R602.10.3(1)
Bracing for Winds

TABLE R602.10.3(3)
Seismic Wall Bracing

TABLE R602.10.3(4)
Adjustment Factors – Seismic

R602.10.6.5
Stone and Masonry Veneer

R609.4.1
Garage Doors

R702.7
Vapor Retarders

R703.2, R703.7.3
Water-Resistive Barriers

TABLE R703.8.4(1)
Veneer Attachment

R703.11.2
Vinyl Siding Installation Over Foam Plastic Sheathing

R704
Soffits

R802
Wood Roof Framing

TABLE R802.5.2(1)
Heel Joint Connections

R802.6
Rafter and Ceiling Joist Bearing

TABLE R804.3
CFS Roof Framing Fasteners

R905.4.4.1
Metal Roof Shingle Wind Resistance

Table R403.1(1)
Footings Below Light-Frame Construction

CHANGE TYPE: Modification

CHANGE SUMMARY: Tables R403.1(1), (2) and (3) are revised to more accurately reflect current practice.

2021 CODE TEXT:

TABLE R403.1(1) Minimum Width and Thickness for Concrete Footings for Light-Frame Construction (inches) [a, b, c, d]

Ground Snow Load or Roof Live Load	Story and Type of Structure with Light Frame	Load Bearing Value of Soil (psf)					
		1500	2000	2500	3000	3500	4000
20 psf Roof Live Load or 25 psf Ground Snow Load	1 story - slab on grade	12 × 6	12 × 6	12 × 6	12 × 6	12 × 6	12 × 6
	1 story - with crawl space	12 × 6	12 × 6	12 × 6	12 × 6	12 × 6	12 × 6
	1 story - plus basement	16 × 6 / ~~18 × 6~~	12 × 6 / ~~14 × 6~~	12 × 6	12 × 6	12 × 6	12 × 6
	2 story - slab on grade	13 × 6 / ~~12 × 6~~	12 × 6	12 × 6	12 × 6	12 × 6	12 × 6
	2 story - with crawl space	15 × 6 / ~~16 × 6~~	12 × 6	12 × 6	12 × 6	12 × 6	12 × 6
	2 story - plus basement	19 × 6 / ~~22 × 6~~	14 × 6 / ~~16 × 6~~	12 × 6 / ~~13 × 6~~	12 × 6	12 × 6	12 × 6
	3 story - slab on grade	16 × 6 / ~~14 × 6~~	12 × 6	12 × 6	12 × 6	12 × 6	12 × 6
	3 story - with crawl space	18 × 6 / ~~19 × 6~~	14 × 6	12 × 6	12 × 6	12 × 6	12 × 6
	3 story - plus basement	22 × 7 / ~~25 × 8~~	16 × 6 / ~~19 × 6~~	13 × 6 / ~~15 × 6~~	12 × 6 / ~~13 × 6~~	12 × 6	12 × 6
30 psf	1 story - slab on grade	12 × 6	12 × 6	12 × 6	12 × 6	12 × 6	12 × 6
	1 story - with crawl space	13 × 6	12 × 6	12 × 6	12 × 6	12 × 6	12 × 6
	1 story - plus basement	16 × 6 / ~~19 × 6~~	12 × 6 / ~~14 × 6~~	12 × 6	12 × 6	12 × 6	12 × 6
	2 story - slab on grade	13 × 6 / ~~12 × 6~~	12 × 6	12 × 6	12 × 6	12 × 6	12 × 6
	2 story - with crawl space	16 × 6 / ~~17 × 6~~	12 × 6 / ~~13 × 6~~	12 × 6	12 × 6	12 × 6	12 × 6
	2 story - plus basement	19 × 6 / ~~23 × 6~~	14 × 6 / ~~17 × 6~~	12 × 6 / ~~14 × 6~~	12 × 6	12 × 6	12 × 6
	3 story - slab on grade	16 × 6 / ~~15 × 6~~	14 × 6 / ~~12 × 6~~	12 × 6	12 × 6	12 × 6	12 × 6
	3 story - with crawl space	19 × 6 / ~~20 × 6~~	14 × 6 / ~~15 × 6~~	12 × 6	12 × 6	12 × 6	12 × 6
	3 story - plus basement	22 × 7 / ~~26 × 8~~	16 × 6 / ~~20 × 6~~	13 × 6 / ~~16 × 6~~	12 × 6 / ~~13 × 6~~	12 × 6	12 × 6

(*continues*)

TABLE R403.1(1) (continued)

Ground Snow Load or Roof Live Load	Story and Type of Structure with Light Frame	Load Bearing Value of Soil (psf)					
		1500	2000	2500	3000	3500	4000
50 psf	1 story - slab on grade	12 × 6	12 × 6	12 × 6	12 × 6	12 × 6	12 × 6
	1 story - with crawl space	<u>14 × 6</u> ~~16 × 6~~	12 × 6	12 × 6	12 × 6	12 × 6	12 × 6
	1 story - plus basement	<u>18 × 6</u> ~~21 × 6~~	<u>13 × 6</u> ~~16 × 6~~	<u>12 × 6</u> ~~13 × 6~~	12 × 6	12 × 6	12 × 6
	2 story - slab on grade	<u>15 × 6</u> ~~14 × 6~~	<u>13 × 6</u> ~~12 × 6~~	12 × 6	12 × 6	12 × 6	12 × 6
	2 story - with crawl space	<u>17 × 6</u> ~~19 × 6~~	<u>13 × 6</u> ~~14 × 6~~	12 × 6	12 × 6	12 × 6	12 × 6
	2 story - plus basement	<u>21 × 7</u> ~~25 × 7~~	<u>15 × 6</u> ~~19 × 6~~	<u>12 × 6</u> ~~15 × 6~~	12 × 6	12 × 6	12 × 6
	3 story - slab on grade	<u>18 × 6</u> ~~17 × 6~~	13 × 6	12 × 6	12 × 6	12 × 6	12 × 6
	3 story - with crawl space	<u>20 × 6</u> ~~22 × 6~~	<u>15 × 6</u> ~~17 × 6~~	<u>12 × 6</u> ~~13 × 6~~	12 × 6	12 × 6	12 × 6
	3 story - plus basement	<u>24 × 8</u> ~~28 × 9~~	<u>18 × 6</u> ~~21 × 6~~	<u>14 × 6</u> ~~17 × 6~~	<u>12 × 6</u> ~~14 × 6~~	12 × 6	12 × 6
70 psf	1 story - slab on grade	<u>14 × 6</u> ~~12 × 6~~	12 × 6	12 × 6	12 × 6	12 × 6	12 × 6
	1 story - with crawl space	<u>16 × 6</u> ~~18 × 6~~	<u>12 × 6</u> ~~13 × 6~~	12 × 6	12 × 6	12 × 6	12 × 6
	1 story - plus basement	<u>19 × 6</u> ~~24 × 7~~	<u>14 × 6</u> ~~18 × 6~~	<u>12 × 6</u> ~~14 × 6~~	12 × 6	12 × 6	12 × 6
	2 story - slab on grade	<u>17 × 6</u> ~~16 × 6~~	12 × 6	12 × 6	12 × 6	12 × 6	12 × 6
	2 story - with crawl space	<u>19 × 6</u> ~~21 × 6~~	<u>14 × 6</u> ~~16 × 6~~	<u>12 × 6</u> ~~13 × 6~~	12 × 6	12 × 6	12 × 6
	2 story - plus basement	<u>22 × 7</u> ~~27 × 9~~	<u>17 × 6</u> ~~20 × 6~~	<u>13 × 6</u> ~~16 × 6~~	<u>12 × 6</u> ~~14 × 6~~	12 × 6	12 × 6
	3 story - slab on grade	<u>20 × 6</u> ~~19 × 6~~	<u>15 × 6</u> ~~14 × 6~~	12 × 6	12 × 6	12 × 6	12 × 6
	3 story - with crawl space	<u>22 × 7</u> ~~25 × 7~~	<u>16 × 6</u> ~~18 × 6~~	<u>13 × 6</u> ~~15 × 6~~	12 × 6	12 × 6	12 × 6
	3 story - plus basement	<u>24 × 8</u> ~~30 × 10~~	<u>19 × 6</u> ~~23 × 6~~	<u>15 × 6</u> ~~18 × 6~~	<u>13 × 6</u> ~~15 × 6~~	<u>12 × 6</u> ~~13 × 6~~	12 × 6

a. ~~Interpolation allowed.~~ <u>Linear interpolation of footing width is permitted between the soil bearing pressures in the table.</u> Extrapolation is not ~~allowed~~ <u>permitted</u>.
b. ~~Based on 32-foot-wide house with load-bearing center wall that carries half of the tributary attic, and floor framing. For every 2 feet of adjustment to the width of the house, add or subtract 2 inches of footing width and 1 inch of footing thickness (but not less than 6 inches thick).~~ <u>The table is based on the following conditions and loads: Building width: 32 feet; Wall height: 9 feet; Basement wall height: 8 feet; Dead loads: 15 psf roof and ceiling assembly, 10 psf floor assembly, 12 psf wall assembly; Live loads: Roof and ground snow loads as listed, 40 psf first floor, 30 psf second and third floors. Footing sizes are calculated assuming a clear span roof/ceiling assembly and an interior bearing wall or beam at each floor.</u>
c. <u>Where the building width perpendicular to the wall footing is greater than 32 feet, the footing width shall be increased by 2 inches and footing depth shall be increased by 1 inch for every 4 feet of increase in building width.</u>

(continues)

TABLE R403.1(1) (continued)

d. <u>Where the building width perpendicular to the wall footing is less than 32 feet, a 2 inch decrease in footing width and 1 inch decrease in footing depth is permitted for every 4 feet of decrease in building width, provided the minimum width is 12 inches (304.8 mm) and minimum depth is 6 inches (152.4 mm).</u>

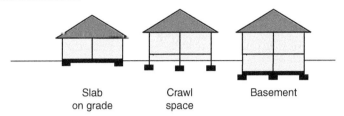

Slab on grade   Crawl space   Basement

TABLE R403.1(2) Minimum Width and Thickness for Concrete Footings for Light-Frame Construction with Brick Veneer <u>or Lath And Plaster</u> (inches)[a, b, c, d]

GROUND SNOW LOAD OR ROOF LIVE LOAD	STORY AND TYPE OF STRUCTURE WITH BRICK VENEER	LOAD-BEARING VALUE OF SOIL (psf)					
		1500	2000	2500	3000	3500	4000

TABLE R403.1(3) Minimum Width and Thickness for Concrete Footings with Cast-In-Place Concrete or ~~Fully~~ <u>Partially</u> Grouted Masonry Wall Construction (inches)[a, b, c, d]

GROUND SNOW LOAD OR ROOF LIVE LOAD	STORY AND TYPE OF STRUCTURE WITH CMU OR CONCRETE	LOAD-BEARING VALUE OF SOIL (psf)					
		1500	2000	2500	3000	3500	4000

(For changes to Tables R403.1(2) and R403.1(3) please see the 2021 IRC)

CHANGE SIGNIFICANCE: Designers using Tables R403.1(1), (2) or (3), minimum width and thickness for concrete footings, introduced in the 2015 *International Residential Code* (IRC), have found in certain instances footing widths required by the table to be different than those required by previous editions of the IRC. In fact, due to conservative assumptions for the tables, some footing widths were wider than an engineering analysis

Stem wall and interior footings.

would suggest necessary. A review of underlying calculations found minimum widths and thicknesses where load assumptions were incorrect. Therefore, changes to the tables were proposed for the *2021 International Residential Code*.

Revised assumptions and calculations for concrete footing tables include the following:

1. Application of roof snow load rather than ground snow load to the roof. The actual roof snow load per ASCE 7, unadjusted by any other factors, is 70 percent of the ground snow load or 20 pounds per square foot, whichever is greater. Consistent with Chapter 8 rafter tables, a thermal factor (C_t) of 1.1 per ASCE 7 is also applied to roof snow load calculations.

2. A 100-pound per square foot (psf) load was previously used for above-grade concrete or masonry walls, representing a solid or fully grouted 8-inch CMU wall. Such walls are more likely to be either 8-inch CMU grouted at 48 inches on center or 8-inch insulated concrete forms, both of which impose only a 55 pound per square foot load. This change affects Table R403.1(3) for footings under above-grade concrete or CMU walls.

3. Previous calculations used the ASCE 7 load combination applying a 0.75 factor for concurrent roof/snow and floor live loads, ignoring load combinations that apply to just a roof/attic live load, just a snow load, or just a floor live load. These additional load combinations apply for a single-story building and for interior footings.

4. Calculations were formerly based on tributary width, yet footnote b added 2 inches of footing width for every 2 feet of additional building width. For a building with interior concrete footings, the tributary width is half the distance between footings, not half the entire building width. As a result of confusing building and tributary width, footnote b potentially doubled the additional footing width for buildings wider than 32 feet.

In many cases, revised footing widths in the 2021 IRC are more consistent with historic practice, while still technically justified under engineering standards and accepted practices. There are a few cases for houses on weaker soils (1500 psf and 2000 psf soil bearing strength) with slab-on-grade or crawlspace foundations, where a revised assumption of clear-spanning roof trusses led to a slight increase to footing widths.

Footnote b allowing adjustment of footing width and depth is now divided into two footnotes. Footnote c requires an increase in footing width and depth when the building width perpendicular to a wall footing exceeds 32 feet. Footnote d permits, but does not require, a decrease in footing width and depth for a building width narrower than 32 feet.

Example: Footing Size with Variable Building Width

A single-family home has the following attributes:

- Two-story
- Gable roof
- Crawl space foundation
- Thirty psf ground snow load

- 1500 psi soil capacity assumed
- Building width varies between 12 and 32 feet
- Clear-span trusses at 24 inches on center
- Center-bearing floors

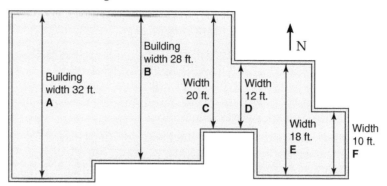

For building widths A – F, the tributary length is:

A. Tributary length = 16 ft
B. Tributary length = 14 ft
C. Tributary length = 10 ft
D. Tributary length = 6 ft
E. Tributary length = 9 ft
F. Tributary length = 5 ft

For building widths A–F, the worst case footing size is a 16 × 6 footing for a 32-foot building width. The footing could be reduced to 12 × 6 where applicable on the north and south walls of the building.

Footings support building loads which include the weight of the building (dead loads), live loads (people and furnishings) and environmental loads, for example snow. Exterior bearing walls carry roof loads based on an assumption of a clear-span roof, such as a truss, which means the roof's tributary area is calculated based on half the building width. With an assumption of a center-bearing wall or beam carrying the load from floor joists or trusses, the tributary area for floors is based on one-quarter of the building width.

Gable-end exterior walls carry only the weight of nonload-bearing walls and the roof load from one-half of the truss or rafter spacing. In the example, the truss spacing is 24 inches on-center so the roof load tributary width is only 1 foot. Gable-end walls will typically never need more than the 12 × 6 minimum footing size.

For a hip roof, it is reasonable to measure the building width perpendicular to the wall from the peak of the hip (where it connects to the ridge) to the end wall—in other words, the length of the longest hip truss or rafter.

R406.2 Foundation Waterproofing

CHANGE TYPE: Deletion

CHANGE SUMMARY: Six-mil polyvinyl chloride and polyethylene fabrics are removed from the list of approved waterproofing materials.

2021 CODE TEXT: R406.2 Concrete and masonry foundation waterproofing. In areas where a high water table or other severe soil-water conditions are known to exist, exterior foundation walls that retain earth and enclose interior spaces and floors below grade shall be waterproofed from the higher of (a) the top of the footing or (b) 6 inches (152 mm) below the top of the basement floor, to the finished grade. Walls shall be waterproofed in accordance with one of the following:

1. Two-ply hot-mopped felts.
2. Fifty-five-pound (25 kg) roll roofing.
3. ~~Six-mil (0.15 mm) polyvinyl chloride.~~
4. ~~Six-mil (0.15 mm) polyethylene.~~
5. Forty-mil (1 mm) polymer-modified asphalt.
6. Sixty-mil (1.5 mm) flexible polymer cement.

Approved waterproofing material.

7. One-eighth inch (3 mm) cement-based, fiber-reinforced, waterproof coating.
8. Sixty-mil (1.5 mm) solvent-free liquid-applied synthetic rubber.

All joints in membrane waterproofing shall be lapped and sealed with an adhesive compatible with the membrane.

Exception: Organic-solvent-based products such as hydrocarbons, chlorinated hydrocarbons, ketones and esters shall not be used for ICF walls with expanded polystyrene form material. Use of plastic roofing cements, acrylic coatings, latex coatings, mortars and pargings to seal ICF walls is permitted. Cold-setting asphalt or hot asphalt shall conform to Type C of ASTM D449. Hot asphalt shall be applied at a temperature of less than 200°F (93°C).

CHANGE SIGNIFICANCE: Waterproofing is the formation of a durable and impervious barrier designed to prevent water from entering a specific building envelope section, generally the foundation. To be effective, a waterproofing system consists of durable and continuous materials applied to all foundation areas that will be subjected to ground water. During backfill, materials containing debris, frost, sharp stones or rocks may scrape along the foundation tearing thinner waterproofing materials.

Section R406.2 is amended by deleting Items 3 and 4 from the list of approved products that can be used as a waterproofing material. Six-mil polyvinyl chloride and polyethylene products have not proven to possess the thickness or strength to be effective and durable in this application. These materials are likely to rip or tear allowing water behind the waterproofing material, trapping moisture behind undamaged fabric and potentially creating issues for the building interior.

Other products approved as waterproofing materials are made of heavier materials more resistant to damage or displacement during backfill, with minimum 1/8-inch or 40-mil (approximately 3/64-inch) thickness.

R408.8 Vapor Retarder in Crawlspaces

CHANGE TYPE: Addition

CHANGE SUMMARY: Where exposed to grade in a crawl space, a Class I or II vapor retarder is required on exposed air permeable insulation between floor joists in Climate Zones 1A, 2A and 3A.

2021 CODE TEXT: <u>**R408.8 Under-floor vapor retarder.** In Climate Zones 1A, 2A, and 3A below the warm-humid line, a continuous Class I or II vapor retarder shall be provided on the exposed face of air permeable insulation installed between the floor joists and exposed to the grade in the under-floor space. The vapor retarder shall have a maximum water vapor permeance of 1.5 perms when tested in accordance with Procedure B of ASTM E96.</u>

<u>**Exception:** The vapor retarder shall not be required in unvented crawl spaces constructed in accordance with Section R408.3.</u>

CHANGE SIGNIFICANCE: New Section R408.8 addresses issues with moisture accumulation in floors above vented crawl spaces in warm-humid climates. Water vapor migrating from vented crawl spaces or post and beam foundations toward cooler and drier indoor spaces may cause mold, mildew and decay within floor assemblies, especially where

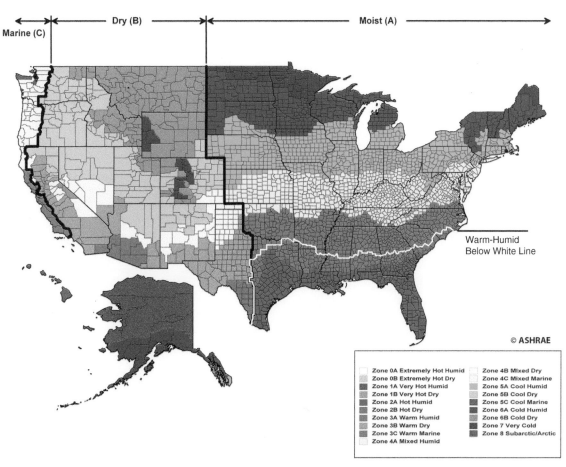

Locations where vapor retarder is required under floor joists.

an impermeable floor covering or underlayment is used, as moisture can condense and be trapped within the wood subfloor. Such moisture problems have been observed even where crawl spaces are constructed in accordance with the IRC, with properly sized and located ventilation openings.

This section requires a Class I or Class II vapor retarder on the exposed face of air-permeable insulation materials installed between floor framing members above the crawl space. The vapor retarder can be a separate layer of material or incorporated as part of the insulation. Examples include foil facing on fiberglass batts, polyisocyanurate rigid foam, or a 6-mil polyethylene sheet applied over permeable insulation along the base of floor joists, I-joists or trusses.

R506.2.3 Vapor Retarders Under Concrete Slabs

CHANGE TYPE: Modification

CHANGE SUMMARY: Thicker vapor retarders are now required below slabs-on-grade.

2021 CODE TEXT: R506.2.3 Vapor retarder. A <u>minimum</u> ~~6-mil~~ <u>10-mil</u> (~~0.006~~ <u>0.010</u> inch; ~~152 μm~~ <u>0.254 mm</u>) ~~polyethylene or approved~~ vapor retarder <u>conforming to ASTM E1745 Class A requirements</u> with joints lapped not less than 6 inches (152 mm) shall be placed between the concrete floor slab and the base course or the prepared subgrade where a base course does not exist.

> **Exception:** The vapor retarder is not required for the following:
> 1. Garages, utility buildings and other unheated accessory structures.
> 2. For unheated storage rooms having an area of less than 70 square feet (6.5 m) and carports.
> 3. Driveways, walks, patios and other flatwork not likely to be enclosed and heated at a later date.
> 4. Where approved by the building official, based on local site conditions.

CHANGE SIGNIFICANCE: Water vapor migrating from the ground into spaces such as vented crawlspaces and open foundation systems or through concrete slabs on ground toward cooler and drier indoor spaces may cause mold, mildew and decay, as condensed moisture is trapped indoors. Thin membranes on the ground can be torn during construction allowing moisture to migrate up into the house.

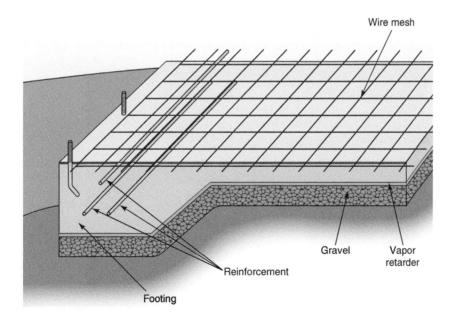

Vapor retarder laid between subgrade and concrete.

Section R506.2.3 requires a vapor retarder between a concrete slab and the top of underlying soil or gravel. The vapor retarder is now required to be a minimum of 10-mil thickness and may be any material that meets ASTM E1745 Class A requirements. The greater thickness offers increased resistance to moisture transmission and provides increased durability during and after installation.

The updated minimum vapor retarder requirements now meet American Concrete Institute (ACI) recommendations as well. ACI 302.1, *Guide to Concrete Floor and Slab Construction*, requires a below slab vapor retarder to meet the requirements of ASTM E1745, *Standard Specification for Plastic Water Vapor Retarders Used in Contact with Soil or Granular Fill under Concrete Slabs*.

Significant Changes to the IRC 2021 Edition R507 ■ Deck Loads

R507
Deck Loads

CHANGE TYPE: Modification

CHANGE SUMMARY: Deck design is now based on live and snow loads.

2021 CODE TEXT: R507.1 Decks. Wood-framed decks shall be in accordance with this section. <u>Decks shall be designed for the live load required in Section R301.5 or the ground snow load indicated in Table R301.2, whichever is greater.</u> For decks using materials and conditions not prescribed in this section, refer to Section R301.

CHANGE SIGNIFICANCE: *International Residential Code* (IRC) prescriptive deck provisions historically have only assumed a 40 psf live load and 10 psf dead load for all components in the deck. However, a significant portion of the population in the United States lives in areas where the ground snow load exceeds the live load in Table R301.5, Minimum Uniformly Distributed Live Loads.

For the 2021 IRC, a deck is now either designed for a 40 pounds per square foot (psf) live load or for the ground snow load listed in a jurisdiction's table of climatic and geographic design criteria (Table R301.2). This

Snow on deck.

requires use of whichever load is higher. Updated lumber tables consider ground snow loads of 50, 60 and 70 psf while allowing interpolation between loads.

For snow loading, an increase in wood strength is accounted for using a load duration factor from the *National Design Specification (NDS) for Wood Construction*. While deck geometry and nearby structures can cause drifting, these effects are outside the scope of IRC deck tables. Similarly, elevated decks have snow loads less than the ground snow load based on ASCE 7, but this reduction is not included to provide simpler tables.

Note that when comparing the *2021 International Building Code* (IBC) and the 2021 IRC, minimum deck live loads will be 1.5 times the associated room live load per the 2021 IBC. For a sleeping room, this will be 1.5 × 30 psf or 45 psf. For all other residential areas, the deck live load will be 1.5 × 40 psf or 60 psf. In the IRC, for decks accessed from any room, the minimum live load remains 40 psf.

R507.3 Deck Footings

CHANGE TYPE: Modification

CHANGE SUMMARY: Clarifications are made for freestanding deck footing exceptions and a tributary area of 5 psf is added to the deck footing size table.

2021 CODE TEXT: R507.3 Footings. Decks shall be supported on concrete footings or other approved structural systems designed to accommodate all loads in accordance with Section R301. Deck footings shall be sized to carry the imposed loads from the deck structure to the ground as shown in Figure R507.3. ~~The footing depth shall be in accordance with Section R403.1.4.~~

~~Exception~~ <u>Exceptions:</u>

1. <u>Footings shall not be required for free</u> ~~Free~~-standing decks consisting of joists directly supported on grade over their entire length.
2. <u>Footings shall not be required for freestanding decks that meet all of the following criteria:</u>
   2.1. <u>The joists bear directly on precast concrete pier blocks at grade without support by beams or posts,</u>
   2.2. <u>The area of the deck does not exceed 200 square feet (18.6 m²).</u>

Footings for buried pressure-treated deck posts.

2.3. The walking surface is not more than 20 inches (508 mm) above grade at any point within 36 inches (914 mm) measured horizontally from the edge.

TABLE R507.3.1 Minimum Footing Size for Decks

		SOIL BEARING CAPACITY[a,c,d]								
		1500[e] psf			2000[e] psf			≥ 3000[e] psf		
LIVE OR GROUND SNOW LOAD[b] (psf)	TRIBUTARY AREA (sq. ft.)	Side of a square footing (inches)	Diameter of a round footing (inches)	Thickness[f] (inches)	Side of a square footing (inches)	Diameter of a round footing (inches)	Thickness[f] (inches)	Side of a square footing (inches)	Diameter of a round footing (inches)	Thickness[f] (inches)
40	5	7	8	6	7	8	6	7	8	6
	20	10	12	6	9	9	6	7	8	6
	40	14	16	6	12	14	6	10	12	6
50	5	7	8	6	7	8	6	7	8	6
	20	11	13	6	10	11	6	8	9	6
	40	15	17	6	13	15	6	11	13	6
60	5	7	8	6	7	8	6	7	8	6
	20	12	14	6	11	12	6	9	10	6
70	5	7	8	6	7	8	6	7	8	6
	20	12	14	6	11	13	6	9	10	6

For SI: 1 inch = 25.4 mm, 1 square foot = 0.0929 m^2, 1 pound per square foot = 0.0479 kPa.

a. Interpolation permitted; extrapolation not permitted.
b. Based on highest load case: Dead + Live or Dead + Snow.
c. ~~Assumes minimum square footing to be 12 inches × 12 inches × 6 inches for 6 × 6 post.~~ Footing dimensions shall allow complete bearing of the post.
d. If the support is a brick or CMU pier, the footing shall have a minimum 2-inch projection on all sides.
e. Area, in square feet, of deck surface supported by post and footings.
f. Minimum thickness shall only apply to plain concrete footings.

(Deleted text not shown for clarity: values in tables that did not change are not shown for brevity.)

CHANGE SIGNIFICANCE: Footing exceptions for freestanding decks have been clarified. In the first exception, joists supported on grade do not require footings. In the second exception, a small deck low to the ground may use concrete piers as footings directly supporting deck joists (no beams or posts). The exception limits these decks to 200 square feet and a height to the top of the deck of 20 inches above grade.

Table R507.3.1, minimum footing size for decks, is expanded to offer a minimum footing size decreased from a 12-inch by 12-inch square in the 2018 IRC to a smaller 7-inch by 7-inch square or 8-inch round footing in the 2021 IRC based on a new 5 psf tributary area. The former limit of a 12-inch by 12-inch footing was significantly oversized for small areas such as stairs or landings. The smaller tributary area also allows for some precast concrete solutions for small landings and porches.

Updated footnote c clarifies that the footing must be wide enough to allow complete bearing of the post. New footnote f states that minimum footing thickness is based on plain concrete. A thinner reinforced footing may be possible with calculations.

R507.4 Deck Posts

CHANGE TYPE: Modification

CHANGE SUMMARY: The deck post height table is expanded by adding the tributary area supported by a post and the wood species for determination of maximum post height.

2021 CODE TEXT: R507.4 Deck posts. For single-level ~~wood-framed~~ decks, ~~with beams sized in accordance with Table R507.5, deck~~ wood post size shall be in accordance with Table R507.4.

TABLE R507.4 Deck Post Height

Loads[b] (psf)	Post Species[c]	Post Size[d]	Tributary Area[g,h] (sq.ft.)							
			20	40	60	80	100	120	140	160
			Maximum Deck Post Height[a] (feet-inches)							
40 Live Load	Southern Pine	4 × 4	14-0	13-8	11-0	9-5	8-4	7-5	6-9	6-2
		4 × 6	14-0	14-0	13-11	12-0	10-8	9-8	8-10	8-2
		6 × 6	14-0	14-0	14-0	14-0	14-0	14-0	14-0	14-0
		8 × 8	14-0	14-0	14-0	14-0	14-0	14-0	14-0	14-0
	Douglas Fir[e], Hem-fir[e], SPF[e]	4 × 4	14-0	13-6	10-10	9-3	8-0	7-0	6-2	5-3
		4 × 6	14-0	14-0	13-10	11-10	10-6	9-5	8-7	7-10
		6 × 6	14-0	14-0	14-0	14-0	14-0	14-0	14-0	14-0
		8 × 8	14-0	14-0	14-0	14-0	14-0	14-0	14-0	14-0
	Redwood[f], Western Cedars[f], Ponderosa Pine[f], Red Pine[f]	4 × 4	14-0	13-2	10-3	8-1	5-8	NP	NP	NP
		4 × 6	14-0	14-0	13-6	11-4	9-9	8-4	6-9	4-7
		6 × 6	14-0	14-0	14-0	14-0	14-0	14-0	13-7	9-7
		8 × 8	14-0	14-0	14-0	14-0	14-0	14-0	14-0	14-0
50 Ground Snow Load	Southern Pine	4 × 4	14-0	12-2	9-10	8-5	7-5	6-7	5-11	5-4
		4 × 6	14-0	14-0	12-6	10-9	9-6	8-7	7-10	7-3
		6 × 6	14-0	14-0	14-0	14-0	14-0	14-0	14-0	13-4
		8 × 8	14-0	14-0	14-0	14-0	14-0	14-0	14-0	14-0
	Douglas Fir[e], Hem-fir[e], SPF[e]	4 × 4	14-0	12-1	9-8	8-2	7-1	6-2	5-3	4-2
		4 × 6	14-0	14-0	12-4	10-7	9-4	8-4	7-7	6-11
		6 × 6	14-0	14-0	14-0	14-0	14-0	14-0	14-0	12-10
		8 × 8	14-0	14-0	14-0	14-0	14-0	14-0	14-0	14-0
	Redwood[f], Western Cedars[f], Ponderosa Pine[f], Red Pine[f]	4 × 4	14-0	11-8	9-0	6-10	3-7	NP	NP	NP
		4 × 6	14-0	14-0	12-0	10-0	8-6	7-0	5-3	NP
		6 × 6	14-0	14-0	14-0	14-0	14-0	14-0	10-8	2-4
		8 × 8	14-0	14-0	14-0	14-0	14-0	14-0	14-0	14-0

(continues)

TABLE R507.4 (*continued*)

Loads[b] (psf)	Post Species[c]	Post Size[d]	Tributary Area[g,h] (sq/ft.)							
			20	40	60	80	100	120	140	160
			Maximum Deck Post Height[a] (feet-inches)							
60 Ground Snow Load	Southern Pine	4 × 4	14-0	11-1	8-11	7-7	6-7	5-10	5-2	4-6
		4 × 6	14-0	14-0	11-4	9-9	8-7	7-9	7-1	6-6
		6 × 6	14-0	14-0	14-0	14-0	14-0	14-0	12-9	11-2
		8 × 8	14-0	14-0	14-0	14-0	14-0	14-0	14-0	14-0
	Douglas Fir[e], Hem-fir[e], SPF[e]	4 × 4	14-0	10-11	8-8	7-3	6-2	5-0	3-7	NP
		4 × 6	14-0	13-11	11-2	9-7	8-4	7-5	6-8	5-11
		6 × 6	14-0	14-0	14-0	14-0	14-0	14-0	12-2	10-2
		8 × 8	14-0	14-0	14-0	14-0	14-0	14-0	14-0	14-0
	Redwood[f], Western Cedars[f], Ponderosa Pine[f], Red Pine[f]	4 × 4	14-0	10-6	7-9	4-7	NP	NP	NP	NP
		4 × 6	14-0	13-7	10-9	8-9	7-0	4-9	NP	NP
		6 × 6	14-0	14-0	14-0	14-0	14-0	9-9	NP	NP
		8 × 8	14-0	14-0	14-0	14-0	14-0	14-0	14-0	14-0
70 Ground Snow Load	Southern Pine	4 × 4	14-0	10-2	8-2	6-11	5-11	5-2	4-4	3-4
		4 × 6	14-0	12-11	10-5	8-11	7-10	7-1	6-5	5-10
		6 × 6	14-0	14-0	14-0	14-0	14-0	12-9	10-11	8-7
		8 × 8	14-0	14-0	14-0	14-0	14-0	14-0	14-0	14-0
	Douglas Fir[e], Hem-fir[e], SPF[e]	4 × 4	14-0	10-1	7-11	6-6	5-3	3-7	NP	NP
		4 × 6	14-0	12-10	10-3	8-9	7-7	6-8	5-10	4-11
		6 × 6	14-0	14-0	14-0	14-0	14-0	12-2	9-9	5-9
		8 × 8	14-0	14-0	14-0	14-0	14-0	14-0	14-0	14-0
	Redwood[f], Western Cedars[f], Ponderosa Pine[f], Red Pine[f]	4 × 4	14-0	9-5	6-5	NP	NP	NP	NP	NP
		4 × 6	14-0	12-6	9-8	7-7	5-3	NP	NP	NP
		6 × 6	14-0	14-0	14-0	14-0	10-8	NP	NP	NP
		8 × 8	14-0	14-0	14-0	14-0	14-0	14-0	14-0	14-0

For SI: 1 inch = 25.4 mm, 1 foot = 304.8 mm, 1 pound per square foot = 0.0479 kPa., NP = Not Permitted
a. Measured from the underside of the beam to the top of footing or pier.
b. 10 psf dead load. Snow load not assumed to be concurrent with live load.
c. No. 2 grade, wet service factor included.
d. Notched deck posts shall be sized to accommodate beam size in accordance with Section R507.5.2
e. Includes incising factor.
f. Incising factor not included.
g. Area, in square feet, of deck surface supported by post and footings.
h. Interpolation permitted. Extrapolation not permitted.

(*Deleted text not shown for clarity.*)

CHANGE SIGNIFICANCE: Table R507.4, Maximum Deck Post Height, is greatly expanded to allow for a larger variety of post heights. By including consideration of tributary area supported by a post for smaller decks, porches and landings, the table allows greater heights with 4 × 4 and 4 × 6 posts. Having a maximum tributary area also defines an upper size limit for decks unless additional posts are added. Similar to deck footing size provisions, consideration for 50, 60 and 70 psf ground snow loads is

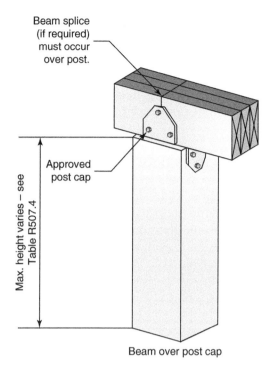

Deck post maximum height varies.

provided. All tabulated wood species can attain the maximum post height of 14 feet with an 8 × 8 minimum nominal dimension or a maximum tributary area of 20 square feet. Southern pine, Douglas Fir, Hem-fir and Spruce-Pine-Fir posts with 6 × 6 minimum nominal dimensions can also achieve the 14-foot maximum for decks carrying a 40 psf live load.

Standard wood assumptions are shown in the footnotes. Lumber is assumed to be wet, of No. 2 grade and carry a 10 psf dead load. Some species require incising to provide appropriate preservative treatment retention levels.

R507.5
Deck Beams

CHANGE TYPE: Modification

CHANGE SUMMARY: The deck beam span table is split into multiple tables providing spans for given deck live or snow loads. Single and multi-ply spans as well as options for support of cantilevered deck joists are listed.

2021 CODE TEXT: **R507.5 Deck Beams.** Maximum allowable spans for wood deck beams, as shown in Figure R507.5, shall be in accordance with ~~Table R507.5.~~ Tables R507.5(1) through R507(4). Beam plies shall be fastened together with two rows of 10d (3-inch × 0.128-inch) nails minimum at 16 inches (406 mm) on center along each edge. Beams shall be permitted to cantilever at each end up to one-fourth of the ~~allowable~~ actual beam span. Deck beams of other materials shall be permitted where designed in accordance with accepted engineering practices.

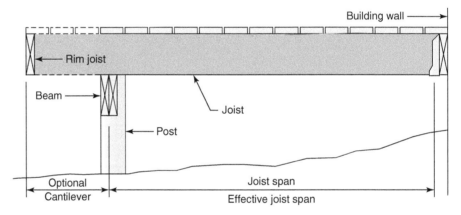

Deck beam span.

TABLE R507.5(2) Maximum Deck Beam Span - 50 PSF Ground Snow Load[c]

Beam Species[d]	Beam Size[e]	Effective Deck Joist Span Length[a,i,j] (feet)						
		6	8	10	12	14	16	18
		Maximum Beam Span[a,b,f] (feet-inches)						
Southern Pine	1-2 × 6	4-6	3-11	3-6	3-2	2-11	2-9	2-7
	1-2 × 8	5-9	4-11	4-5	4-0	3-9	3-6	3-3
	1-2 × 10	6-9	5-10	5-3	4-9	4-5	4-2	3-11
	1-2 × 12	8-0	6-11	6-2	5-8	5-3	4-11	4-7
	2-2 × 6	6-8	5-9	5-2	4-9	4-4	4-1	3-10
	2-2 × 8	8-6	7-4	6-7	6-0	5-7	5-2	4-11
	2-2 × 10	10-1	8-9	7-10	7-1	6-7	6-2	5-10
	2-2 × 12	11-11	10-3	9-2	8-5	7-9	7-3	6-10
	3-2 × 6	7-11	7-2	6-6	5-11	5-6	5-1	4-10
	3-2 × 8	10-5	9-3	8-3	7-6	6-11	6-6	6-2
	3-2 × 10	12-8	10-11	9-9	8-11	8-3	7-9	7-3
	3-2 × 12	14-11	12-11	11-6	10-6	9-9	9-1	8-7

(continues)

TABLE R507.5(2) (continued)

Beam Species[d]	Beam Size[e]	Effective Deck Joist Span Length[a,i,j] (feet)						
		6	8	10	12	14	16	18
		Maximum Beam Span[a,b,f] (feet-inches)						
Douglas fir-larch[g], Hem-fir[g], Spruce-pine-fir[g]	1-2 × 6	4-0	3-5	2-11	2-7	2-4	2-2	2-0
	1-2 × 8	5-4	4-7	3-11	3-5	3-1	2-10	2-8
	1-2 × 10	6-7	5-8	4-11	4-5	4-0	3-8	3-5
	1-2 × 12	7-7	6-7	5-11	5-4	4-10	4-6	4-2
	2-2 × 6	6-0	5-2	4-7	4-2	3-10	3-5	3-2
	2-2 × 8	8-0	6-11	6-2	5-8	5-0	4-7	4-2
	2-2 × 10	9-9	8-5	7-7	6-11	6-4	5-10	5-4
	2-2 × 12	11-4	9-10	8-9	8-0	7-5	6-11	6-6
	3-2 × 6	7-6	6-6	5-9	5-3	4-11	4-7	4-4
	3-2 × 8	10-0	8-8	7-9	7-1	6-6	6-1	5-8
	3-2 × 10	12-3	10-7	9-6	8-8	8-0	7-6	7-0
	3-2 × 12	14-3	12-4	11-0	10-1	9-4	8-9	8-3
Redwood[h], Western cedars[h], Ponderosa pine[h], Red pine[h]	1-2 × 6	4-1	3-6	3-0	2-8	2-5	2-3	2-1
	1-2 × 8	5-2	4-6	4-0	3-6	3-2	2-11	2-9
	1-2 × 10	6-4	5-6	4-11	4-6	4-1	3-9	3-6
	1-2 × 12	7-4	6-4	5-8	5-2	4-10	4-6	4-3
	2-2 × 6	6-1	5-3	4-8	4-4	3-11	3-6	3-3
	2-2 × 8	7-8	6-8	5-11	5-5	5-0	4-8	4-3
	2-2 × 10	9-5	8-2	7-3	6-8	6-2	5-9	5-5
	2-2 × 12	10-11	9-5	8-5	7-8	7-2	6-8	6-3
	3-2 × 6	7-1	6-5	5-11	5-5	5-0	4-8	4-5
	3-2 × 8	9-4	8-4	7-5	6-10	6-4	5-11	5-7
	3-2 × 10	11-9	10-2	9-1	8-4	7-8	7-2	6-9
	3-2 × 12	13-8	11-10	10-7	9-8	8-11	8-4	7-10

For SI: 1 inch = 25.4 mm, 1 foot = 304.8 mm, 1 pound per square foot = 0.0479 kPa, 1 pound = 0.454 kg.

a. Interpolation allowed. Extrapolation is not allowed.
b. Beams supporting a single span of joists with or without cantilever.
c. Dead load = 10 psf, L/Δ = 360 at main span, L/Δ = 180 at cantilever. Snow load not assumed to be concurrent with live load.
d. No. 2 grade, wet service factor included.
e. Beam depth shall be equal to or greater than the depth of intersecting joist for a flush beam connection.
f. Beam cantilevers are limited to the adjacent beam's span divided by 4.
g. Includes incising factor.
h. Incising factor not included.
i. Deck joist span as shown in Figure R507.5
j. For calculation of effective deck joist span, the actual joist span length shall be multiplied by the joist span factor in accordance with R507.5(5).

TABLE R507.5(5) Joist Span Factors for Calculating Effective Deck Joist Span [for use with footnote j in Tables R507.5(1), R507.5(2), R507.5(3) and R507.5(4)]

C/J[a]	Joist Span Factor
0 (no cantilever)	0.66
1/12 (0.87)	0.72
1/10 (0.10)	0.80
1/8 (0.125)	0.84
1/6 (0.167)	0.90
1/4 (0.250)	1.00

a. C = actual joist cantilever length (feet) J = actual joist span length (feet)

(Deleted table text not shown for clarity; to see Tables R507.5(1), (3) or (4) refer to the 2021 IRC.)

CHANGE SIGNIFICANCE: Similar to changes for deck post heights, there is a need to consider snow loads greater than 40 psf for deck beams. The 2018 IRC's prescriptive deck provisions only included a 40 psf live load and 10 psf dead load. Table R507.5, Deck beam span lengths, is replaced by four tables, R507.5(1) – (4), which account for 50, 60 and 70 psf ground snow load conditions. Single-ply spans are now listed for all wood species while there continue to be multi-ply span options for each wood species.

Maximum beam spans consider the load from tributary areas based on joist spans. In Tables R507.5(1)-(4), all deck joists are assumed to cantilever the allowable one-quarter of the joist back span past the supporting beam. This assumption is included in the calculated tributary area of the beam. When a beam supports a shorter joist cantilever, new Table R507.5(5) may be used to determine an effective joist span. When the joist cantilever is shorter than ¼ of the back span or there is no cantilever, the joist span may be decreased when determining the beam's maximum span. This allows the maximum beam span to increase due to the shorter effective joist span. Several examples are provided to illustrate the effective joist span:

Example 1:
A deck with a ground snow load of 50 psf is designed using two plies of Southern Pine 2 × 10.

Joist span is 12 feet and there is no cantilever.

- C – cantilever, J – joist
- C = 0 feet
- J = 12 feet
- Without footnote j, Table R507.5(2) limits the beam to a maximum span of 7'-1".

Applying the adjustment factor from footnote j and Table R507.5(5):

- C / J = 0 and the joist span factor is 0.66.
- An effective joist span can be calculated as 0.66 × 12' = 8".

- The maximum beam span is 8'-9" per Table R507.5(2) because there is no cantilever.

Note: The beam span is not reduced by 0.66, rather the joist span is reduced by 0.66 to determine an effective joist span.

Example 2:

A deck with a ground snow load of 50 psf is designed using two plies of Southern Pine 2 × 10.

Joist span is 12 feet and there is a 12-inch cantilever.

- C – cantilever, J – joist
- C = 1 feet
- J = 12 feet
- Without footnote j, Table R507.5(2) limits the beam to a maximum span of 7'-1".

Applying the adjustment factor from footnote j and Table R507.5(5):

- C/J = 1/12 therefore the joist span factor is 0.72.
- An effective joist span can be calculated as 0.72 × 12' = 8'-8".
- A maximum beam span can be determined from Table R507.5(2) for a 10' effective joist span = 7'-10".
- Or by interpolating per footnote a, a beam span of 8'-5" is achieved as follows:

Maximum Beam Span   (converted to inches):

	A	B	C
Effective joist span	8 ft. 96 in.	8 ft. 8 in. 104 in.	10 ft. 120 in.
Maximum beam span	105 in.	?	94 in.

To interpolate, subtract the number of inches between the 10 ft (column C) and 8 ft (column A) joist spans, which is 24 inches. The effective joist span in column B is 8 inches longer than the smaller span in column A, which is 8/24 or 1/3 the difference in lengths. Subtract beam span C from beam span A to get 11 inches. Multiply 11 inches by 1/3 to get 3.67 inches. Then take beam span A which is 105 inches and subtract 4 inches (rounded up from 3.67 inches). This gives a maximum beam span of 101" or 8'-5".

Maximum Beam Span

	A	B	C
Effective joist span	96 in.	104 in.	120 in.
Maximum beam span	105 in.	101 in. 8 ft. 5 in.	94 in.

Example 3:

A deck with a ground snow load of 50 psf is designed using two plies of Southern Pine 2 × 10.

Joist span is 12 feet and there is a 3-foot cantilever.

- C – cantilever, J – joist
- C = 3 feet
- J = 12 feet
- Without footnote j, Table R507.5(2) limits the beam to a maximum span of 7'-1".

Applying the adjustment factor from footnote j and Table R507.5(5):

- C/J = 3/12 or 1/4 and the joist span factor is 1.0.

Tabulated beam spans are based on a joist with a cantilever of 1/4 of the back span therefore the maximum beam span remains 7'-1".

R507.6 Deck Joists

CHANGE TYPE: Modification

CHANGE SUMMARY: Deck joist options are added for decks with large ground snow loads. Cantilever spans are now specifically based on maximum joist spans.

2021 CODE TEXT: R507.6 Deck joists. Maximum allowable spans for wood deck joists, as shown in Figure R507.6, shall be in accordance with Table R507.6. The maximum joist spacing shall be limited by the decking materials in accordance with Table R507.7. ~~The maximum joist cantilever shall be limited to one-fourth of the joist span or the maximum cantilever length specified in Table R507.6, whichever is less.~~

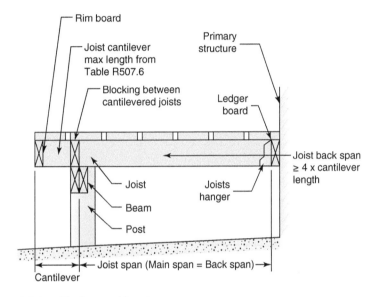

Joists with dropped beam.

TABLE R507.6 Maximum Deck Joist Spans

Load[a] (psf)	Joist Species[b]	Joist Size	Allowable Joist Span[c] (feet-inches)			Maximum Cantilever[d] (feet-inches)						
			Joist Spacing (inches)			Joist Back Span[g] (feet)						
			12	16	24	4	6	8	10	12	14	16	18
40 Live Load	Southern Pine	2 × 6	9-11	9-0	7-7	1-0	1-6	1-5	NP	NP	NP	NP	NP
		2 × 8	13-1	11-10	9-8	1-0	1-6	2-0	2-6	2-3	NP	NP	NP
		2 × 10	16-2	14-0	11-5	1-0	1-6	2-0	2-6	3-0	3-4	3-4	NP
		2 × 12	18-0	16-6	13-6	1-0	1-6	2-0	2-6	3-0	3-6	4-0	4-1
	Douglas Fir[e], Hem-fir[e], Spruce-Pine-Fir[e]	2 × 6	9-6	8-4	6-10	1-0	1-6	1-4	NP	NP	NP	NP	NP
		2 × 8	12-6	11-1	9-1	1-0	1-6	2-0	2-3	2-0	NP	NP	NP
		2 × 10	15-8	13-7	11-1	1-0	1-6	2-0	2-6	3-0	3-3	NP	NP
		2 × 12	18-0	15-9	12-10	1-0	1-6	2-0	2-6	3-0	3-6	3-11	3-11
	Redwood[f], Western Cedars[f], Ponderosa Pine[f], Red Pine[f]	2 × 6	8-10	8-0	6-10	1-0	1-4	1-1	NP	NP	NP	NP	NP
		2 × 8	11-8	10-7	8-8	1-0	1-6	2-0	1-11	NP	NP	NP	NP
		2 × 10	14-11	13-0	10-7	1-0	1-6	2-0	2-6	3-0	2-9	NP	NP
		2 × 12	17-5	15-1	12-4	1-0	1-6	2-0	2-6	3-0	3-6	3-8	NP

(*continues*)

TABLE R507.6 (*continued*)

Load[a] (psf)	Joist Species[b]	Joist Size	Allowable Joist Span[c] (feet-inches) Joist Spacing (inches)			Maximum Cantilever[d] (feet-inches) Joist Back Span[g] (feet)							
			12	16	24	4	6	8	10	12	14	16	18
50 Ground Snow Load	Southern Pine	2 × 6	9-2	8-4	7-4	1-0	1-6	1-5	NP	NP	NP	NP	NP
		2 × 8	12-1	11-0	9-5	1-0	1-6	2-0	2-5	2-3	NP	NP	NP
		2 × 10	15-5	13-9	11-3	1-0	1-6	2-0	2-6	3-0	3-1	NP	NP
		2 × 12	18-0	16-2	13-2	1-0	1-6	2-0	2-6	3-0	3-6	3-10	3-10
	Douglas Fir[e], Hem-fir[e], Spruce-Pine-Fir[e]	2 × 6	8-10	8-0	6-8	1-0	1-6	1-4	NP	NP	NP	NP	NP
		2 × 8	11-7	10-7	8-11	1-0	1-6	2-0	2-3	NP	NP	NP	NP
		2 × 10	14-10	13-3	10-10	1-0	1-6	2-0	2-6	3-0	3-0	NP	NP
		2 × 12	17-9	15-5	12-7	1-0	1-6	2-0	2-6	3-0	3-6	3-8	NP
	Redwood[f], Western Cedars[f], Ponderosa Pine[f], Red Pine[f]	2 × 6	8-3	7-6	6-6	1-0	1-4	1-1	NP	NP	NP	NP	NP
		2 × 8	10-10	9-10	8-6	1-0	1-6	2-0	1-11	NP	NP	NP	NP
		2 × 10	13-10	12-7	10-5	1-0	1-6	2-0	2-6	2-9	NP	NP	NP
		2 × 12	16-10	14-9	12-1	1-0	1-6	2-0	2-6	3-0	3-5	3-5	NP
60 Ground Snow Load	Southern Pine	2 × 6	8-8	7-10	6-10	1-0	1-6	1-5	NP	NP	NP	NP	NP
		2 × 8	11-5	10-4	8-9	1-0	1-6	2-0	2-4	NP	NP	NP	NP
		2 × 10	14-7	12-9	10-5	1-0	1-6	2-0	2-6	2-11	2-11	NP	NP
		2 × 12	17-3	15-0	12-3	1-0	1-6	2-0	2-6	3-0	3-6	3-7	NP
	Douglas Fir[e], Hem-fir[e], Spruce-Pine-Fir[e]	2 × 6	8-4	7-6	6-2	1-0	1-6	1-4	NP	NP	NP	NP	NP
		2 × 8	10-11	9-11	8-3	1-0	1-6	2-0	2-2	NP	NP	NP	NP
		2 × 10	13-11	12-4	10-0	1-0	1-6	2-0	2-6	2-10	NP	NP	NP
		2 × 12	16-6	14-3	11-8	1-0	1-6	2-0	2-6	3-0	3-5	3-5	NP
	Redwood[f], Western Cedars[f], Ponderosa Pine[f], Red Pine[f]	2 × 6	7-9	7-0	6-2	1-0	1-4	NP	NP	NP	NP	NP	NP
		2 × 8	10-2	9-3	7-11	1-0	1-6	2-0	1-11	NP	NP	NP	NP
		2 × 10	13-0	11-9	9-7	1-0	1-6	2-0	2-6	2-7	NP	NP	NP
		2 × 12	15-9	13-8	11-2	1-0	1-6	2-0	2-6	3-0	3-2	NP	NP
70 Ground Snow Load	Southern Pine	2 × 6	8-3	7-6	6-5	1-0	1-6	1-5	NP	NP	NP	NP	NP
		2 × 8	10-10	9-10	8-2	1-0	1-6	2-0	2-2	NP	NP	NP	NP
		2 × 10	13-9	11-11	9-9	1-0	1-6	2-0	2-6	2-9	NP	NP	NP
		2 × 12	16-2	14-0	11-5	1-0	1-6	2-0	2-6	3-0	3-5	3-5	NP
	Douglas Fir[e], Hem-fir[e], Spruce-Pine-Fir[e]	2 × 6	7-11	7-1	5-9	1-0	1-6	NP	NP	NP	NP	NP	NP
		2 × 8	10-5	9-5	7-8	1-0	1-6	2-0	2-1	NP	NP	NP	NP
		2 × 10	13-3	11-6	9-5	1-0	1-6	2-0	2-6	2-8	NP	NP	NP
		2 × 12	15-5	13-4	10-11	1-0	1-6	2-0	2-6	3-0	3-3	NP	NP
	Redwood[f], Western Cedars[f], Ponderosa Pine[f], Red Pine[f]	2 × 6	7-4	6-8	5-10	1-0	1-4	NP	NP	NP	NP	NP	NP
		2 × 8	9-8	8-10	7-4	1-0	1-6	1-11	NP	NP	NP	NP	NP
		2 × 10	12-4	11-0	9-0	1-0	1-6	2-0	2-6	2-6	NP	NP	NP
		2 × 12	14-9	12-9	10-5	1-0	1-6	2-0	2-6	3-0	3-0	NP	NP

For SI: 1 inch = 25.4 mm, 1 foot = 304.8 mm, 1 pound per square foot = 0.0479 kPa, 1 pound = 0.454 kg. NP = Not Permitted

a. Dead load = 10 psf. Snow load not assumed to be concurrent with live load.
b. No. 2 grade, wet service factor included.
c. $L/\Delta = 360$ at main span.
d. $L/\Delta = 180$ at cantilever with 220-pound point load applied to end.
e. Includes incising factor.
f. Incising factor not included.
g. Interpolation permitted. Extrapolation is not permitted.

(*Deleted text not shown for clarity.*)

CHANGE SIGNIFICANCE: Table R507.6, maximum deck joist spans, has created confusion for determining cantilever lengths for given joist spans. The 2018 IRC table listed an allowable cantilever length in terms of joist spacing. Since the assumed main span was the maximum allowable joist span for that spacing, the maximum cantilevers in some cases were not intuitive. Determining whether a joist and its cantilever met the maximum limits was time consuming and unclear. The second cantilever limit of one-quarter of the main span was located within the footnotes where it may have been overlooked. In the 2018 IRC, designers had to check both the table's maximum cantilever length and the limit of not more than one-fourth of the actual joist span.

With the updated 2021 IRC table, cantilevers are based on the actual back span creating a more intuitive maximum cantilever length. The term back span is used to differentiate the actual joist span from the maximum joist span available in the table. Where the table states a cantilever is not permitted, the back span in the table is longer than the maximum permitted joist span. An example of this is a 2 × 6 joist which has a maximum joist span in the 6.5-foot to 9-foot range. The table has entries for back spans of 10 feet to 18 feet but the cantilever length for a 2 × 6 is shown as "Not Permitted" since a joist span of 10 feet is not permitted for a 2 × 6.

New table organization now first lists maximum joist spans based on their on-center spacing. Then the table lists maximum cantilever lengths. Cantilevers, in order to carry their loads, must have the joist extend back along the deck a distance longer than the cantilever's length. Generally, a back span needs to be at least four times as long as the cantilever.

For example, for a Southern Pine 2 × 6 joist, if the joist spans 8 feet from the ledger to a supporting beam, then the joist can continue past the beam and cantilever up to 18 inches beyond the beam. But a Southern Pine 2 × 8 joist can cantilever 2 feet or a full one-quarter of the joist back span (distance between ledger and beam).

Table R507.6 is also updated to account for ground snow loads of 50, 60 and 70 pounds per square foot (psf).

R507.7
Decking

CHANGE TYPE: Modification

CHANGE SUMMARY: The wood decking table is updated to show maximum on-center joist spacing for single- and multi-span configurations.

2021 CODE TEXT: R507.7 Decking. Maximum allowable spacing for joists supporting <u>wood</u> decking, <u>excluding stairways,</u> shall be in accordance with Table R507.7. Wood decking shall be attached to each supporting member with not less than two 8d threaded nails or two No. 8 wood screws. <u>Maximum allowable spacing for joists supporting plastic composite decking shall be in accordance with Section R507.2.</u> Other approved decking or fastener systems shall be installed in accordance with the manufacturer's installation requirements.

TABLE R507.7 Maximum Joist Spacing for <u>Wood</u> Decking

Decking Material Type and Nominal Size	~~Maximum On-Center Joist Spacing[c]~~			
	Decking perpendicular to joist		Decking diagonal to joist[a]	
	<u>Single Span[c]</u>	Multi-Span[c]	<u>Single Span[c]</u>	Multi-Span[c]
	<u>Maximum On-Center Joist Spacing (inches)</u>			
1¼ inch-thick wood deck boards[b]	<u>12</u>	16 ~~inches~~	<u>8</u>	<u>12</u> ~~inches~~
2-inch-thick wood	<u>24</u>	24 ~~inches~~	<u>18</u>	<u>24</u> ~~16~~ inches
~~Plastic composite~~	~~In accordance with Section R507.2~~		~~In accordance with Section R507.2~~	

For SI: 1 inch = 25.4 mm, 1 foot = 304.8 mm, 1 degree = 0.01745 rad.
a. Maximum angle of 45 degrees from perpendicular for wood deck boards.
b. <u>Or other maximum span provided by an accredited lumber grading or inspection agency.</u>
c. <u>Individual wood deck boards supported by two joists shall be considered single span and three or more joists shall be considered multi-span.</u>

Decking may span two or more joists.

CHANGE SIGNIFICANCE: Table R507.7, Maximum Joist Spacing for Decking, is conservative for 2-inch nominal wood decking material. When evaluated using the American Lumber Standard Committee's (ALSC) decking policy, 2-inch nominal material can span 24 inches rather than 16 inches as shown in the 2018 IRC. Similarly, 5/4-inch decking is rated per ALSC's decking policy. While the minimum rated span is 16 inches, for certain species and grades, the allowable joist spacing may increase. The 2021 IRC Table R507.7 keeps a conservative baseline while increasing flexibility for manufacturers and designers.

The maximum joist spacing is updated to consider whether decking spans across two joists (called a single span) or three or more joists (a multi-span condition). Load resistance changes when decking is laid across multiple joists. This can change the allowable joist spacing. A multispan condition with decking allows an increase in the maximum joist spacing in some cases, particularly when decking is relatively thin and flexible. If a wider spacing of joists is desired for 5/4-inch-thick wood decking, footnote b allows a lumber grading agency or third-party wood inspection agency to provide maximum spans for specific 5/4-inch-thick decking.

R507.10
Exterior Guards

CHANGE TYPE: Addition

CHANGE SUMMARY: Specific requirements for deck guards are added.

2021 CODE TEXT:

R507.10 Exterior guards. Guards shall be constructed to meet the requirements of Sections R301.5 and R312 and this section.

R507.10.1 Support of guards. Where guards are supported on deck framing, guard loads shall be transferred to the deck framing with a continuous load path to the deck joists.

R507.10.1.1 Guards supported by side of deck framing. Where guards are connected to the interior or exterior side of a deck joist or beam, the joist or beam shall be connected to the adjacent joists to prevent rotation of the joist or beam. Connections relying only on fasteners in end grain withdrawal are not permitted.

R507.10.1.2 Guards supported on top of deck framing. Where guards are mounted on top of the decking, the guards shall be connected to the deck framing or blocking and installed in accordance with manufacturer's instructions to transfer the guard loads to the adjacent joists.

R507.10.2 Wood posts at deck guards. Where 4-inch by 4-inch (102 mm by 102 mm) wood posts support guard loads applied to the top of the guard, such posts shall not be notched at the connection to the supporting structure.

Deck guards are used for fall protection.

R507.10.3 Plastic composite guards. Plastic composite guards shall comply with the provisions of Section R507.2.2.

R507.10.4 Other guards. Other guards shall be in accordance with manufacturer's instructions or in accordance with accepted engineering principles.

CHANGE SIGNIFICANCE: The 2018 IRC had no requirements for constructing exterior guards on decks in Section R507. Guards provide the first line of defense against significant falls, which can result in serious and sometimes fatal injuries. Exterior guards on decks, particularly the guard system connection to the deck framing, are rarely engineered and even more rarely tested to verify adequacy to meet the 200-pound load requirements of Table R301.5, Minimum Live Loads.

Exterior deck guards must continue to meet Section R312 requirements and the loads listed in Table R301.5. The new provisions also reinforce the need for a load path from the guard and rail into the deck joists, beams or blocking to which a guard is connected. End grain connections in withdrawal are prohibited. In other words, guard fasteners may not be installed into the ends of deck joists or beams if loading will occur parallel to the length of the joist or beam slowly pulling the fasteners out of the lumber. When guards are connected to the side of beams or joists, the beam or joist shall be connected to adjacent joists—for example by blocking or straps—to resist rotation of the beam or joist when load is applied to the guard.

Table R602.3(1)
Fasteners – Roof and Wall

CHANGE TYPE: Modification

CHANGE SUMMARY: Additional fastener options are added to the fastener table for roof and walls.

2021 CODE TEXT:

EXCERPT OF TABLE R602.3(1) Fastening Schedule

Item	Description of Building Elements	Number and Type of Fastener[a,b,c,e]	Spacing and Location
	Roof		
1	Blocking between ceiling joists, or rafters or trusses to top plate or other framing below	4-8d box (2½" × 0.113") nails 3-8d common (2½" × 0.131") nails 3-10d box (3" × 0.128") nails 3-(3" × 0.131") nails	Toenail
	Blocking between rafters or truss not at the wall top plates, to rafter or truss	2-8d common (2½" × 0.131") nails 2-(3" × 0.131") nails	Each end, toenail
		2-16d common (3½" × 0.162") nails 3-(3" × 0.131") nails	End nail
	Flat blocking to truss and web filler	16d common (3½" × 0.162") nails 3-(3" × 0.131") nails	6" o.c. face nail
	Wall		
12	Adjacent full-height stud to end of header	3-16d common (3½" × 0.162") nails 4-16d box (3 ½" × 0.135") nails 4-10d box (3" × 0.128") nails 4-(3" × 0.131") nails	End nail

CHANGE SIGNIFICANCE: *International Residential Code* (IRC) Table R602.3(1) and *International Building Code* (IBC) Table 2304.10.1 are essentially the same table. Although the fastener tables are closely aligned, there are variations in the prescribed fasteners within the two tables. Some fastener options are offered only in the IRC table and other options are offered in the IBC table and not the IRC table. Changes to the 2021 codes attempt to harmonize fastener options in the two tables. Item 1 of the table shows such changes with addition of nailing options and clarifying text describing blocking attachment to specific roof elements.

King studs nailed to header.

The required nailing of the first full-height stud adjacent to a header is added to the fastening schedule table in Item 12 for wall connections, where it can be located quickly. Additional full-height studs are fastened to each other in accordance with Item 8 of Table R602.3(1), Stud to stud (not at braced wall panels) or Item 9, Stud to stud and abutting studs at intersecting wall corners (at braced wall panels).

Table R602.3(1)
Fasteners – Roof Sheathing

CHANGE TYPE: Modification

CHANGE SUMMARY: Additional fastener options are added to the fastener table in the roof sheathing section while maximum field nailing is reduced.

2021 CODE TEXT:

EXCERPT OF TABLE R602.3(1) Fastening Schedule

Item	Description of Building Elements	Number and Type of Fastener[a,b,c]	Spacing of Fasteners Edges[h] (inches)	Intermediate supports[c,e] (inches)
	Wood structural panels (WSP), subfloor, roof and interior wall sheathing to framing and particleboard wall sheathing to framing				
~~30~~ 31	$3/8"$ – $1/2"$	6d common or deformed (2" × 0.113" × 0.266" head)-~~(subfloor, wall)~~[i]; $2^3/8"$ × 0.113" × 0.266" head nail (subfloor, wall)[i]	6[f]	~~12~~ 6[f]
		8d common (2½" × 0.131") (roof) RSRS-01 ($2^3/8"$ × 0.113") nail (roof)[ib]	6[f]	~~12~~ 6[f]
~~31~~ 32	$19/32"$ – ~~1"~~ $3/4"$	8d common (2 – 2½" × 0.131") (subfloor, wall) Deformed $2^3/8"$ × 0.113" × 0.266" head (wall or subfloor)	6	12
		8d common (2½" × 0.131") nail (roof) RSRS-01 ($2^3/8"$ × 0.113") nail (roof)[ib]	6[f]	~~12~~ 6[f]
~~32~~ 33	~~1-1/8"~~ $7/8"$ – $1^1/4"$	10d common (3" × 0.148") nail ~~8d~~ (2½" × 0.131" × 0.281" head) deformed nail	6	12

For SI: 1 inch = 25.4 mm, 1 foot = 304.8 mm, 1 mile per hour = 0.447 m/s; 1 ksi = 6.895 MPa.

f. For wood structural panel roof sheathing attached to gable end roof framing and to intermediate supports within 48 inches of roof edges and ridges, nails shall be spaced at ~~6~~ 4 inches on center where the ultimate design wind speed is ~~less than 130 mph and shall be spaced 4 inches on center where the ultimate design wind speed is 130 mph or greater but less than 140 mph~~ greater than 130 mph in Exposure B or greater than 110 mph in Exposure C.

CHANGE SIGNIFICANCE: IRC Table R602.3(1) and IBC Table 2304.10.1 are essentially the same table in terms of structural connections. Although the connections are closely aligned, there are variations in the prescribed fasteners within the two tables. Some fastener options are offered only in the IRC table and not in the IBC table and other options are offered in the IBC table and not the IRC table. These changes harmonize fastener options in the two tables. In addition, where additional information exists in one table and not the other, it is added.

Roof sheathing nailing is updated in Table R602.3(1) based on wind load values from ASCE 7, *Minimum Design Loads and Associated Criteria for Buildings and Other Structures*. The table changes are also consistent with the roof sheathing nailing requirements in the *2018 Wood Frame Construction Manual* (WFCM). Wind uplift nailing requirements for

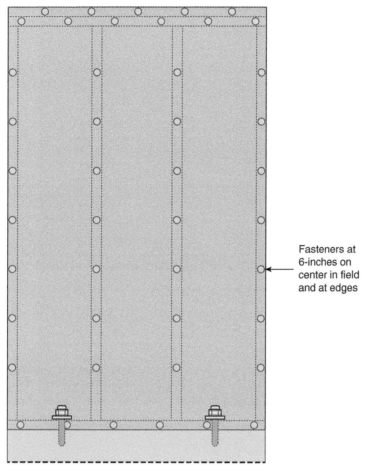

Method WSP fasteners are now required to be a maximum of 6 inches on center.

common species of roof framing with specific gravities of 0.42 or greater, for example SPF and Hem-Fir, are the basis of the new nail spacing requirements in Table R602.3(1) intended to be easy to specify for roof nailing.

The basic roof sheathing nailing schedule is 6 inches on center at panel edges and 6 inches on center at intermediate supports of the panel. As shown in WFCM Table 3.10A for the common case of roof framing spaced at 24 inches on center, nailing at intermediate supports in the interior portions of the roof is 6 inches on center for wind speeds within the scope of IRC. The 6 inches on center spacing is also appropriate for edge zones except where ultimate wind speeds equal or exceed 130 mph in Exposure B and 110 mph in Exposure C where nailing at 4 inches on center for panel edges and intermediate supports is required. This special case is addressed by the modification to footnote f.

Table R602.3(2)
Alternate Attachments

CHANGE TYPE: Clarification

CHANGE SUMMARY: Table R602.3(2) footnote g is updated for clarity.

2021 CODE TEXT:

TABLE R602.3(2) Alternate Attachments to Table R602.3(1)

Nominal Material Thickness (inches)	Description[a,b] of Fastener and Length (inches)	Spacing[c] of Fasteners	
		Edges (inches)	Intermediate supports (inches)
Wood structural panels subfloor, roof[g] and wall sheathing to framing and particleboard wall sheathing to framing[f]			

Nominal Material Thickness (inches)	Description[a,b] of Fastener and Length (inches)	Spacing[c] of Fasteners	
		Edges (inches)	Body of panel[d] (inches)

b. Staples shall have a minimum crown width of $^7/_{16}$-inch ~~on diameter~~ except as noted.

g. ~~Specified alternate attachments for roof sheathing shall be permitted where~~ <u>Alternate fastening is only permitted for roof sheathing where</u> the ultimate design wind speed is less than <u>or equal to 110 mph, and where fasteners are installed 3 inches on center at all supports</u> ~~130 mph. Fasteners attaching wood structural panel roof sheathing to gable end wall framing shall be installed using the spacing listed for panel edges~~.

(No changes to table values)

CHANGE SIGNIFICANCE: Footnote g is updated in Table R602.3(2), Alternate attachments, based on uplift load requirements in ASCE 7, *Minimum Design Loads and Associated Criteria for Buildings and Other Structure* for the alternative fasteners. To keep specification of roof sheathing to relatively simple attachment schedules, reference calculations use a 0.099-inch and 0.113-inch diameter nail at 3-inch on center spacing at all locations. This value is based on the nail shank withdrawal capacity in wood framing with a specific gravity equal to 0.42 (e.g., SPF lumber) and pre-calculated wind uplift loads from Table 3.10 of the *Wood Frame Construction Manual* (WFCM). The use of 3-inch spacing at all supports was extended to staples based on the assumption that the ASCE 7-16 load increase would similarly require reduced spacing. This assumption was applied to staples because a withdrawal value is not available for staples in the *National Design Specification* (NDS) *for Wood Construction*.

Significant Changes to the IRC 2021 Edition   Table R602.3(2) ■ Alternate Attachments   **117**

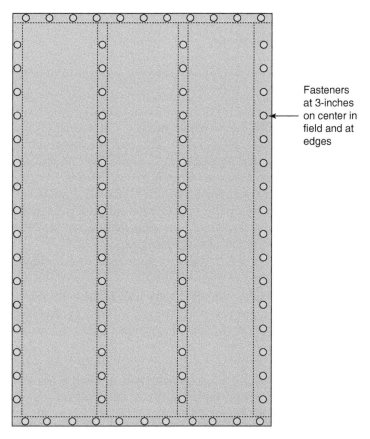

Fasteners at 3-inches on center in field and at edges

Fasteners are now required to be 3 inches on center on roof sheathing.

Therefore, with this change a narrower nail or a staple used in roof sheathing must be placed using a maximum 3-inch on center spacing. These fasteners are only allowed in regions with a wind speed less than or equal to 110 mph. In areas with higher wind speeds, the alternate attachment table may not be used for roof sheathing.

R602.9
Cripple Walls

CHANGE TYPE: Clarification

CHANGE SUMMARY: Cripple wall requirements apply only to exterior cripple walls.

2021 CODE TEXT: R602.9 Cripple walls. Foundation cripple walls shall be framed of studs not smaller than the studding above. Where exceeding 4 feet (1219 mm) in height, such walls shall be framed of studs having the size required for an additional story. ~~Cripple~~ Exterior cripple walls with a stud height less than 14 inches (356 mm) shall be continuously sheathed on one side with wood structural panels fastened to both the top and bottom plates in accordance with Table R602.3(1), or the cripple walls shall be constructed of solid blocking.

Cripple walls shall be supported on continuous foundations.

CHANGE SIGNIFICANCE: The IRC and IBC require foundation cripple walls, below exterior walls, with studs less than 14 inches tall to be "continuously-sheathed" in all seismic regions. This requirement is not related to wall bracing which is covered in Section R602.10. The requirement for continuous sheathing on cripple wall studs with a height less than 14 inches (or solid-blocking) is intended to ensure the integrity of the studs when nails are end-nailed into top and bottom plates by face-nailing sheathing into the top and bottom plates as well as studs.

Cripple wall along exterior of house – unsheathed.

Cripple wall along exterior of house – sheathed.

In regions with shallow frost-depth it is common to have shallow crawl spaces. Typically, a concrete stem wall forms the exterior foundation walls and directly supports the floor. The interior walls, typically 2 to 4 feet tall, are cripple walls laid on a strip footing. These walls move with the exterior concrete walls during an earthquake and have few issues. Therefore, continuous blocking or sheathing is not required for these interior walls when the exterior foundation is concrete up to the floor framing and bottom plate. Continuous sheathing on these short walls in a crawl space also creates issues for ventilation, under-floor mechanical systems, plumbing and access.

Interior cripple wall.

In past earthquakes, exterior cripple walls have been very vulnerable to out-of-plane movement with the cripple wall losing its ability to support the walls above causing the cripple wall to rock out of plumb and collapse. As exterior cripple walls need protection and interior cripple walls simply need to be nailed appropriately, the requirement for a cripple wall to be continuously sheathed is updated from a provision for all cripple walls to a requirement for exterior walls only in the 2021 IRC.

R602.10.1.2

Location of Braced Wall Lines

CHANGE TYPE: Modification

CHANGE SUMMARY: Section R602.10.1.2 limits placement of a braced wall line.

2021 CODE TEXT: **R602.10.1.2** ~~Offsets along a~~ **Location of braced wall lines and permitted offsets.** <u>Each braced wall line shall be located such that no more than two-thirds of the required braced wall panel length is located to one side of the braced wall line. Braced wall panels shall be permitted to be offset up to 4 feet (1219 mm) from the designated braced wall line. Braced wall panels parallel to a braced wall line shall be offset not more than 4 feet (1219 mm) from the designated braced wall line location as shown in Figure R602.10.1.1.</u>

Exterior walls parallel to a braced wall line shall be offset not more than 4 feet (1219 mm) from the designated braced wall line location as shown in Figure R602.10.1.1.

Interior walls used as bracing shall be offset not more than 4 feet (1219 mm) from a braced wall line through the interior of the building as shown in Figure R602.10.1.1.

CHANGE SIGNIFICANCE: Over a series of code cycles, changes to IRC Section R602.10 wall bracing provisions have caused some of the important concepts fundamental to the development of the bracing provisions to be lost. In the 2006 IRC and earlier editions, braced wall panels were required on exterior walls with additional interior braced wall lines where needed to meet braced wall line spacing requirements. The concept that exterior walls are to be braced is not specifically stated in the 2009 IRC forward. Rather, a line is drawn on plans with braced wall panels on walls counted as part of a braced wall line when the panels are within 4 feet of the line drawn on the plans.

This sounds reasonable. It allows the designer to break up the exterior walls pushing some out and others inward along the front of a building. But what about when the front of a house is one single continuous wall? Can the designer still draw the braced wall line 4 feet inward of the actual wall?

The IRC did not address this issue leaving each jurisdiction to decide and designers arguing their case with each jurisdiction. In fact, most jurisdictions feel that the braced wall line must be on a physical wall when the wall line forms a single unbroken line.

For the 2021 edition, the IRC requires that at least one-third of all braced wall panels be either on a braced wall line or on the opposite side of the braced wall line from the other braced wall panels. Braced wall panels continue to be required to be within 4 feet of the braced wall line. For the case where a single wall forms the entire braced wall line, all braced wall panels must be on the braced wall line. In other words, the braced wall line must be drawn at the physical wall.

Braced wall line (BWL) examples:

— — — Braced Wall Line
━━━ Braced Wall Panel (BWP)

Example 1:

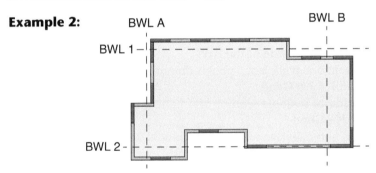

BWL 1: line runs between two walls, 4 of 6 panels on outside side of line, 2 of 6 panels on opposite side of line = OK
BWL 2: line runs between three walls, 1 of 4 panels outside line, 2 of 4 panels on BWL and 1 of 4 panels inside line = OK
BWL A: line runs between two walls, 2 of 3 panels on one side of line, 1 of 3 panels on opposite side of line = OK
BWL B: line runs on one wall = OK

Example 2:

BWL 1: line runs between two walls, 4 of 6 panels on outside side of line, 2 of 6 panels on opposite side of line = OK
BWL 2: line runs between three walls, 1 of 4 panels outside line, 2 of 4 panels on BWL and 1 of 4 panels inside line = OK
BWL A: line runs between two walls, 2 of 3 panels on one side of line, 1 of 3 panels on opposite side of line = OK
BWL B: line runs inside single wall, 2 of 2 panels outside BWL = **No Good**

Example 3:

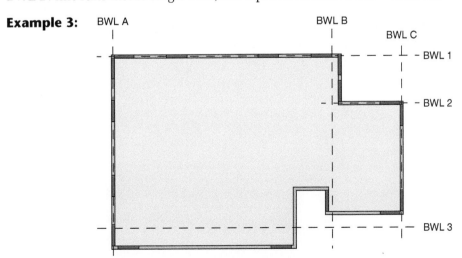

BWL 1: line runs along a single wall, 4 of 4 panels are on the BWL = OK
BWL 2: line runs along a single wall, 1 of 1 panel is on the BWL = OK
BWL 3: line runs between two walls, 2 of 3 panels on one side of line, 1 of 3 panels on opposite side of line = OK
BWL A: line runs along a single wall, 4 of 4 panels are on the BWL = OK
BWL B: line runs between two walls, 2 of 3 panels on one side of line, 1 of 3 panels on opposite side of line = OK
BWL C: line runs along a single wall, 2 of 2 panels are on the BWL = OK

Example 4:

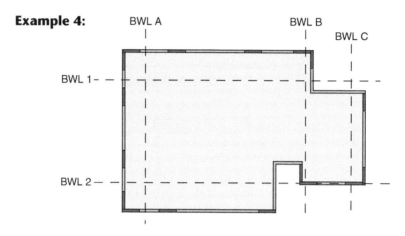

BWL 1: line runs between two walls, 4 of 5 panels on one side of line, 1 of 5 panels on opposite side of line = **No Good**, as more than 2/3 of panels are on one side of the BWL.
BWL 2: line runs on one wall, 3 of 5 panels on one side of line, 2 of 5 panels on the wall = **No Good**
BWL A: line inside of a single wall, 4 of 4 panels are on one side of the BWL = **No Good**
BWL B: line runs between two walls, 1 of 2 panels on one side of line, 1 of 2 panels on opposite side of line = OK
BWL C: line inside of a single wall, 2 of 2 panels are on the inside of the BWL = **No Good**

From the examples, it is clear that when a single wall contains all the braced wall panels in a braced wall line, the BWL must be drawn on the wall. When there are multiple braced wall panels in a BWL, one-third of the panels need to be on one side of the line. This can mean that a long wall with a short wall will need to have the braced wall line placed on the long wall as seen in Example 4's BWL 1. In some cases, the braced wall line will be broken into two separate wall lines like Example 3's BWL 1 and 2.

R602.10.2.2 Location of Braced Wall Panels (BWPs)

CHANGE TYPE: Clarification

CHANGE SUMMARY: Section R602.10.2.2 is clarified for the starting point of the first braced wall panel when not placed at the corner of the structure.

2021 CODE TEXT: **R602.10.2.2 Locations of braced wall panels.** A braced wall panel shall begin within 10 feet (3810 mm) from each end of a braced wall line as determined in Section R602.10.1.1. The distance between adjacent edges of braced wall panels along a braced wall line shall be not greater than 20 feet (6096 mm) as shown in Figure R602.10.2.2.

Exceptions:

1. Braced wall panels in Seismic Design Categories D_0, D_1 and D_2 shall comply with Section R602.10.2.2.1.
2. Braced wall panels with continuous sheathing in Seismic Design Categories A, B or C shall comply with Section R602.10.7.

R602.10.2.2.1 Location of braced wall panels in Seismic Design Categories D_0, D_1 and D_2. Braced wall panels shall be located at each end of a braced wall line.

~~Exception~~ Exceptions:

1. Braced wall panels constructed of Method WSP or BV-WSP and continuous sheathing methods as specified in Section R602.10.4 shall be permitted to begin not more than 10 feet (3048 mm) from each end of a braced wall line provided that each end complies with one of the following:

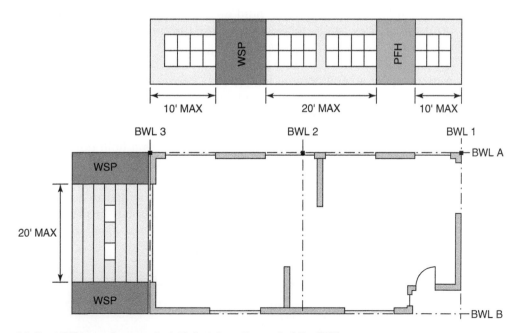

Method PFH panels may start 10 feet from the end of the BWL.

- 1. 1.1. A minimum 24-inch-wide (610 mm) panel for Methods WSP, CS-WSP, CS-G and CSPF is applied to each side of the building corner as shown in End Condition 4 of Figure R602.10.7.
- 2. 1.2. The end of each braced wall panel closest to the end of the braced wall line shall have an 1,000 lb (8 kN) hold-down device fastened to the stud at the edge of the braced wall panel closest to the corner and to the foundation or framing below as shown in End Condition 5 of Figure R602.10.7.

2. Braced wall panels constructed of Method PFH or ABW, or of Method BV-WSP where a hold-down is provided in accordance with Table R602.10.6.5.4, shall be permitted to begin not more than 10 feet from each end of a braced wall line.

CHANGE SIGNIFICANCE: Section R602.10.2.2 deals with placement of braced wall panels on a braced wall line. The main requirement in this section is that the first braced wall panel must begin within 10 feet from the end of the braced wall line. There are two exceptions to this requirement.

1. In Seismic Design Categories (SDC) D_0, D_1 and D_2 additional bracing is required at the corner. Either a hold-down at the BWP or a 2-foot strip of sheathing using wood structural panel at the corner is required.
2. When continuous sheathing methods are used, additional bracing is also required at a corner per Section R602.10.7. Again, either a 2-foot strip of sheathing on each side of the corner or a hold-down at the BWP is installed.

The purpose of the corner sheathing and hold-down options in this section is to restrain the first braced wall panel from overturning, either by having it located at a corner, or by providing a hold-down. When bracing methods already have a hold-down to restrain the panel from overturning, they can be located away from the corner.

Section R602.10.2.2.1 covers bracing in SDC D_0, D_1 and D_2. Braced wall panels may be located up to 10 feet from the corner when the braced wall panel has an 1800-pound hold-down. There are alternate braced wall panels that already have a hold-down of this capacity or higher. Method ABW requires a hold-down with a minimum capacity of 1800 pounds, and Method PFH requires a hold-down with a minimum capacity of 3500 pounds. Generally, Method BW-WSP requires a hold-down with a capacity in excess of 1800 pounds as well. However, there is one case where Method BV-WSP does not require a hold-down, which is considered by adding the text "where a hold-down is provided in accordance with Table R602.10.6.5.4." All three of these methods may be placed up to 10 feet from the end of the braced wall line as the first panel in the wall line.

R602.10.3(1) Bracing for Winds

CHANGE TYPE: Modification

CHANGE SUMMARY: Rows are added to the wind bracing requirements table for 95 mph wind speeds.

2021 CODE TEXT:

TABLE R602.10.3(1) Bracing Requirements Based on Wind Speed

- Exposure Category B
- 30-foot Mean Roof Height
- 10-foot Wall Height
- 2 Braced Wall Lines

Minimum Total Length (feet) of Braced Wall Panels Required Along Each Braced Wall Line[a]

Ultimate Design Wind Speed (mph)	Story Location	Braced Wall Line[c] (feet)	Method LIB[b]	Method GB	Methods DWB, WSP, SFB, PBS, PCP, HPS, BV-WSP, ABW, PFH, PFG, CS-SFB	Methods CS-WSP, CS-G, CS-PF
≤ 95 mph	Top story of 3 / 2 / 1	10	2.5	2.5	1.5	1.5
		20	4.5	4.5	2.5	2.5
		30	6.5	6.5	4.0	3.5
		40	8.5	8.5	5.0	4.0
		50	10.5	10.5	6.0	5.0
		60	12.5	12.5	7.0	6.0
	Middle story of 3 / Bottom of 2	10	5.0	5.0	3.0	2.5
		20	8.5	8.5	5.0	4.5
		30	12.5	12.5	7.0	6.0
		40	16.0	16.0	9.5	8.0
		50	20.0	20.0	11.5	10.0
		60	23.5	23.5	13.5	11.5
	Bottom of 3	10	NP	7.0	4.0	3.5
		20	NP	13.0	7.5	6.5
		30	NP	18.5	10.5	9.0
		40	NP	24.0	13.5	11.5
		50	NP	29.5	17.0	14.5
		60	NP	35.0	20.0	17.0

(*All changes to text or footnotes are shown.*)

CHANGE SIGNIFICANCE: With update of the wind map in Figure R301.2(2), rows for lower wind speeds are needed in Table R602.10.3(1), Bracing requirements based on wind speed. The *Minimum Design Loads and Associated Criteria for Buildings and Other Structures* (ASCE 7-16) wind maps include many areas of the country where wind speeds decrease below 110 mph. This limit was the lower wind speed limit for the western United States in the 2018 IRC. To coordinate the IRC with these new less conservative wind speeds in ASCE 7, existing wind provisions are modified to account for the lower speeds. To see additional explanation about the wind map changes, go to the significant change discussion of Section R301.2.

Wind damage to roof and walls.

For Table R602.10.3(1), bracing requirements based on wind speed, new regions have wind speed as low as 90 mph. The 2018 IRC table began with a wind speed of 110 mph. Continuing to use 110 mph for regions with design wind speeds ranging from 90 to 110 mph is unnecessarily conservative. Because the lowest wind speeds are between 90 and 95 mph, a new category is created. When a design wind speed is less than 95 mph, in other words in western Washington, Oregon or California, the less than 95 mph rows can be used to determine minimum bracing lengths. At this time there are not additional rows for wind speeds less than 100 or 105 mph. Buildings in most areas of Washington, Oregon and California will continue to use 110 mph design wind speed rows in Table R602.10.3(1). Buildings in the mountain states and mid-west will use either 110 or 115 mph design wind speed rows. Areas that are special wind regions will continue to use the wind speeds designated by the state or local jurisdiction.

To identify local wind speeds, the easiest course is to use software currently available from the Applied Technology Council (ATC) or the American Society of Civil Engineers (ASCE). ATC's website is hazards.atcouncil.org and ASCE's website is https://asce7hazardtool.online. Type in an address or GPS coordinates, and it is possible to determine the ground snow load, wind speed, seismic design category and tornado risk in one location. The ATC website remains free to users, while the ASCE website charges a nominal yearly fee and offers wind speeds and tsunami hazard zones for free.

CHANGE TYPE: Clarification

CHANGE SUMMARY: Table R602.10.3(3) labeling and footnotes are updated to clarify use of the table.

Table R602.10.3(3)
Seismic Wall Bracing

2021 CODE TEXT:

TABLE R602.10.3(3) Bracing Requirements Based on Seismic Design Category

- ~~SOIL CLASS D[b]~~
- Wall Height = 10 feet
- 10 PSF Floor Dead Load
- 15 PSF Roof/Ceiling Dead Load
- Braced Wall Line Spacing ≤ 25 feet

Minimum Total Length (Feet) of Braced Wall Panels Required along each Braced Wall Line[a,f,g]

Seismic Design Category[b]	Story Location	Braced Wall Line Length (feet)[c]	Method LIB[d]	Method GB	Methods DWB, SFB, PBS, PCP, HPS, CS-SFB[e]	Methods WSP, PFH[f], PFG[e,f] and ABW[f]	Methods CS-WSP, CS-G, CS-PF
D₂[h]	(upper story)	10	NP	4	4	2.5	2.1
		20	NP	8	8	5	4.3
		30	NP	12	12	7.5	6.4
		40	NP	16	16	10	8.5
		50	NP	20	20	12.5	10.6
	(first story of two)	10	NP	7.5	7.5	5.5	4.7
		20	NP	15	15	11	9.4
		30	NP	22.5	22.5	16.5	14
		40	NP	30	30	22	18.7
		50	NP	37.5	37.5	27.5	23.4
	Cripple wall below one- or two-story dwelling	10	NP	NP	NP	7.5	6.4
		20	NP	NP	NP	15	12.8
		30	NP	NP	NP	22.5	19.1
		40	NP	NP	NP	30	25.5
		50	NP	NP	NP	37.5	31.9

b. ~~Wall bracing lengths are based on a soil site class "D."~~ Interpolation of bracing length between the S_{DS} values associated with the seismic design categories shall be permitted when a site-specific S_{DS} value is determined in accordance with Section 1613.2 of the International Building Code.

<u>f. Methods PFH, PFG and ABW are only permitted on a single story or a first of two stories.</u>

~~f.~~<u>g.</u> Where more than one bracing method is used, mixing methods shall be in accordance with Section R602.10.4.1.

<u>h. One- and two-family dwellings in Seismic Design Category D₂ exceeding two stories shall be designed in accordance with accepted engineering practice.</u>

(*All changes to text or footnotes are shown.*)

CHANGE SIGNIFICANCE: Changes were made to the fundamental structure of the bracing requirements for seismic forces table. The SDCs are no longer based exclusively on Class D soils. This attribute is deleted from the list in the upper left corner of the table. When the site does not have a geotechnical survey, the site soil class may end up based on either Class C or Class D soils. For additional discussion of the changes to soil class see Section R301.2.2 in the *2018 IRC Commentary* or *Significant Changes to the 2018 IRC*.

Damage due to ground movement.

The itemized bracing methods in Table R602.10.3(3) are revised to include all permissible bracing methods. Missing methods included intermittent bracing methods ABW, PFH and PFG; these methods are currently listed in Table R602.10.3(1) for wind forces. Table R602.10.3(3) footnote f is added to identify bracing methods permitted only at the ground floor; in other words, only in single story buildings or on the first floor of two-story buildings. Two of these bracing methods require a hold-down cast in a concrete basement or stem wall or turned down slab edge. Bracing methods with hold-downs limited to the bottom of a two-story building are Methods ABW and PFH. Method PFG is also limited to the bottom of two-stories. Each of these methods was tested in a laboratory to prove its equivalence to Method WSP. In testing, loads used were equivalent to a two-story building's weight. The methods haven't been tested with loads equivalent to a three-story building.

Table R602.10.3(3) footnote h is added to emphasize the two-story height limit for buildings in SDC D_2. If a building is three stories in SDC D_2, an engineered design is required for the building's lateral design. In other words shear walls must be used rather than braced wall panels and diaphragm, beam and column connections are checked for their ability to resist earthquake loads.

Table R602.10.3(4)
Adjustment Factors – Seismic

CHANGE TYPE: Clarification

CHANGE SUMMARY: Table R602.10.3(4) is updated to clarify the limits of brick veneer use and when additional bracing must be used on the building in SDC D_0, D_1 and D_2.

2021 CODE TEXT:

TABLE R602.10.3(4) Seismic Adjustment Factors to the Required Length of Wall Bracing

Item Number	Adjustment Based On	Story[g]	Condition	Adjustment Factor [Multiply length from Table R602.10.3(3) by this factor]	Applicable Methods
7	Walls with stone or masonry veneer, detached one- or two-family dwellings in SDC D_0-D_2	Any Story	See Section R602.10.6.5.4 ~~Table R602.10.6.5~~		BV-WSP
8	Walls with stone or masonry veneer, detached one- or two-family dwellings in SDC D_0-D_2	First and second story of two-story dwelling	Limited Brick Veneer on Second Story. See Section R602.10.6.5.3. ~~Table R602.10.6.5~~	1.2	WSP, CS-WSP
10	Horizontal blocking	Any story	Horizontal blocking omitted	2.0	WSP, PBS, CS-WSP

g. One- and two-family dwellings in Seismic Design Category D_2 exceeding two stories shall be designed in accordance with accepted engineering practice.

(*All changes to text or footnotes are shown.*)

CHANGE SIGNIFICANCE: For wall bracing on dwellings with brick veneer, adjustment factors in Table R602.10.3(4) need to be used. Bracing methods available for use with brick veneer exceeding the first story in high seismic regions include Method BV-WSP, Method WSP and Method CS-WSP.

1. If brick veneer is only on the first story and doesn't extend into gable ends, no additional wall bracing is required—use the minimum bracing length from Table R602.10.3(3).

    Example 1:
    One-story house with bonus room in Seattle, WA

    Brick veneer columns at garage and under front windows only

    Veneer on upper story.

Brick does not extend above the first story, minimum wall bracing length for seismic forces is determined per Table R602.10.3(3) only.

2. If brick veneer extends into the gable-end walls or into an upper story, Method WSP or CS-WSP may be used, if limited brick veneer is placed on the second story. Limited veneer can mean veneer only on the front side of the house at the second story or total veneer area that is less than 25 percent of the second story floor area. With this option, there is a 20 percent increase in the bracing length required by Table R602.10.3(3) using Item 8 of Table R602.10.3(4). Section R602.10.6.5.3 contains the limits for use of these bracing methods.

Example 2:

Two-story house in Portland, OR

Brick veneer across entire front side of building, no veneer on sides or back of building

Veneer above first story on one side of the building.

Brick exists only on the front of the building. Minimum wall bracing length for seismic forces is determined per Table R602.10.3(3) using Method WSP or Method CS-WSP and increasing the required length by 20 percent per Item 8 of Table R602.10.3(4) or use Method BV-WSP.

3. If brick veneer is used to sheathe the full height of multiple second story walls or more than 25 percent of all four second story walls, then Method BV-WSP is the only option. The minimum required length of Method BV-WSP is found in Section R602.10.6.5.4 and Table R602.10.6.5.4.

Example 3:

Two-story house in Memphis, TN

Brick veneer on entire exterior

Brick exists on all sides of the building, minimum wall bracing length for seismic forces is determined per Table R602.10.6.5.4 using Method BV-WSP.

In Table R602.10.3(4), footnote g is added to the Story column to reinforce the two-story limit for brick veneer. A three-story single- or two-family dwelling will require engineering for the lateral system when the building is in SDC D_2. This footnote mirrors the note in Table R602.10.3(3) for clarity.

R602.10.6.5
Stone and Masonry Veneer

CHANGE TYPE: Clarification

CHANGE SUMMARY: Veneer applications in high seismic areas are broken into first story and veneer above the first story applications.

2021 CODE TEXT: R602.10.6.5 Wall bracing for dwellings with stone and masonry veneer in Seismic Design Categories D_0, D_1 and D_2. Townhouses in Seismic Design Categories D_0, D_1 and D_2 with stone or masonry veneer exceeding the first-story height shall be designed in accordance with accepted engineering practice. One- and two-family dwellings in Seismic Design Category D_2 exceeding two stories and having stone or masonry veneer shall be designed in accordance with accepted engineering practice.

Where stone and masonry veneer are installed in accordance with Section R703.8, wall bracing on exterior braced wall lines and braced wall lines on the interior of the building, backing or perpendicular to and laterally supporting veneered walls shall comply with this section.

R602.10.6.5.1 Veneer on First Story Only. Where dwellings in Seismic Design Categories D_0, D_1 and D_2 have stone or masonry veneer installed in accordance with Section R703.8, and the veneer does not exceed the first-story height, wall bracing shall be in accordance with Section R602.10, exclusive of Section R602.10.6.5.

Veneer above first story.

R602.10.6.5.2 Veneer Exceeding First Story Height. Where detached one- or two-family dwellings in Seismic Design Categories D_0, D_1 and D_2 have stone or masonry veneer installed in accordance with Section R703.8, and the veneer exceeds the first-story height, wall bracing at exterior braced wall lines and braced wall lines on the interior of the building shall be constructed using Method BV-WSP in accordance with this section and Figure R602.10.6.5.2. Cripple walls shall not be permitted and required interior braced wall lines shall be supported on continuous foundations.

R602.10.6.5.3 Limited Veneer Exceeding First Story Height. Where detached one- or two-family dwellings in Seismic Design Categories D_0, D_1 and D_2 have exterior veneer installed in accordance with Section R703.8 and brick veneer installed above the first story height meets the following limitations, bracing in accordance with Method WSP or CS-WSP shall be permitted provided that the total length of braced wall panels specified by Table R602.10.3(3) is multiplied by 1.2 for each first- and second-story braced wall line.

1. The dwelling does not extend more than two stories above grade plane.
2. The veneer does not exceed 5 inches (127 mm) in thickness.
3. The height of veneer on gable-end walls does not extend more than 8 feet (2438 mm) above the bearing wall top plate elevation.
4. Where veneer is installed on multiple walls above the first story, the total area of the veneer on the second-story exterior walls shall not exceed 25 percent of the occupied second floor area.
5. Where the veneer is installed on one entire second-story exterior wall, including walls on bay windows and similar appurtenances, brick veneer shall not be installed on any of the other walls on that floor.

R602.10.6.5.4 Length of bracing. The length of bracing along each braced wall line shall be the greater of that required by the ultimate design wind speed and braced wall line spacing in accordance with Table R602.10.3(1) as adjusted by the factors in Table R602.10.3(2) or the seismic design category and braced wall line length in accordance with either Table R602.10.6.5.4 when using Method BV-WSP, or Table R602.10.3(3) as adjusted by the factors in Table R602.10.3(4) when using Method WSP or CS-WSP. Angled walls shall be permitted to be counted in accordance with Section R602.10.1.4, and braced wall panel location shall be in accordance with Section R602.10.2.2. Spacing between braced wall lines shall be in accordance with Table R602.10.1.3. The seismic adjustment factors in Table R602.10.3(4) shall not be applied to the length of bracing determined using Table R602.10.6.5.4, except that the bracing amount increase for braced wall line spacing greater than 25 feet (7620 mm) in accordance with Table R602.10.1.3 shall be required. The minimum total length of bracing in a braced wall line, after all adjustments have been taken, shall be not less than 48 inches (1219 mm) total.

(Deleted text not shown for clarity and brevity)

CHANGE SIGNIFICANCE: Section R602.10.6.5 now begins with two limitations. Townhouses with brick veneer height exceeding the top of the first story require engineering when in SDC D_0, D_1 and D_2. Three-story buildings with brick veneer for one- and two-families in SDC D_2 also require engineering. Note, all three-story one- and two-family dwellings in SDC D_2 require engineering of the lateral force resisting system.

Subsections are now organized so veneer on the building is dealt with in terms of the maximum height and extent of the veneer.

IRC Section	Maximum Veneer Height	Extent Allowed	Bracing Methods Allowed
R602.10.6.5.1	Veneer in first story only	Throughout first story	Any
R602.10.6.5.2	Veneer throughout second story	Throughout first and second stories	Method BV-WSP
R602.10.6.5.3	Veneer in gable or into portion of second story	Throughout first story, limited area in second story	Method WSP, Method CS-WSP, Method BV-WSP

In the 2018 IRC, a section was added to permit limited brick veneer on the second story without triggering the use of Method BV-WSP; meanwhile there has been confusion as to how the provision should be applied. Table R602.10.3(4) is modified to refer to Section R602.10.6.5.3 where the limitations for Methods WSP and CS-WSP reside describing how much brick veneer may extend into a second story. To determine minimum bracing lengths for Methods WSP and CS-WSP, Table R602.10.3(3) is used, adjusted by the 1.2 factor in Table R602.10.3(4). For details on determining whether a building requires Method BV-WSP or may use Methods WSP or CS-WSP, see Significant Change R602.10.3(4).

When brick veneer extends to multiple walls of a second story, Table R602.10.6.5.4 gives the minimum wall bracing required for Method BV-WSP. Method BV-WSP continues to require a hold-down at each end of the braced wall panel from the top story down through the building to the foundation. The method may be used when veneer is only on the first story or when it runs up into a gable end wall; but is only required when all other options are not allowed, in other words, when more than 25 percent of the second story is covered in veneer.

R609.4.1
Garage Doors

CHANGE TYPE: Addition

CHANGE SUMMARY: All garage doors must have a permanent label identifying wind pressure ratings among other information.

2021 CODE TEXT: <u>**R609.4.1 Garage door labeling.** Garage doors shall be labeled with a permanent label provided by the garage door manufacturer. The label shall identify the garage door manufacturer, the garage door model/series number, the positive and negative design wind pressure rating, the installation instruction drawing reference number, and the applicable test standard.</u>

CHANGE SIGNIFICANCE: Since 2005, there has been a push toward considering sustainability in the way our buildings are constructed. If this goal is to be successful, as building owners and occupants increasingly want more information about the sustainability of the buildings they occupy, information to determine how critical components are expected to perform in buildings must be readily available. Some manufacturers already include permanent labels on their products that provide traceability to the manufacturer and the product characteristics.

Garage doors are important components of the building envelope and their performance is critical in preventing wind and water infiltration as well as to maintaining the overall structural integrity of the building. The 2018 IRC did not require garage doors to have a permanent label to provide homeowners, insurers and builders information on a door's wind performance characteristics. For products that do not have permanent labels, it becomes nearly impossible for the owner to determine an installed garage door's structural wind load resistance or energy efficiency. New Section R609.4.1 requires a permanent label on the garage door indicating the manufacturer and model/series number and performance characteristics.

Insurance incentives are now being offered in some states for homes, new and existing, that comply with certain levels of the Fortified program administered by the Insurance Institute for Business and Home Safety (IBHS).

Garage doors must be labeled.

The Fortified program is a set of engineering and building standards designed to help strengthen new and existing homes through system-specific building upgrades to minimum building code requirements that will reduce damage from natural hazards. Fortified offers three different levels of designation depending on the extent of the recommended upgrades to the building's wind resistance. To qualify for a designation, the home is inspected. Without a permanent label indicating the manufacturer and product model/series number, the performance characteristics of garage doors often cannot be determined, and Fortified designations become more difficult to achieve.

R702.7 Vapor Retarders

CHANGE TYPE: Modification

CHANGE SUMMARY: The vapor retarder section is reorganized for clarity and ease of use.

2021 CODE TEXT: R702.7 Vapor retarders. Vapor retarder materials shall be classified in accordance with Table R702.7(1). A vapor retarder shall be provided on the interior side of frame walls of the class indicated in Table R702.7(2), including compliance with Table R702.7(3) or Table R702.7(4) where applicable. An approved design using accepted engineering practice for hygrothermal analysis shall be permitted as an alternative. The climate zone shall be determined in accordance with Section N1101.7.

Exceptions:

1. Basement walls.
2. Below-grade portion of any wall.
3. Construction where accumulation, condensation, or freezing of moisture will not damage the materials.
4. A vapor retarder shall not be required in Climate Zones 1, 2, and 3.

R702.7(1) Vapor Retarder Materials and Classes

CLASS	ACCEPTABLE MATERIALS
I	Sheet polyethylene, nonperforated aluminum foil, or other approved materials with a perm rating of less than or equal to 0.1.
II	Kraft-faced fiberglass batts, vapor retarder paint, or other approved materials applied in accordance with the manufacturer's installation instructions for a perm rating greater than 0.1 and less than or equal to 1.0.
III	Latex paint, enamel paint, or other approved materials applied in accordance with the manufacturer's installation instructions for a perm rating of greater than 1.0 and less than or equal to 10.0.

R702.7(2) Vapor Retarder Options

CLIMATE ZONE	VAPOR RETARDER CLASS		
	CLASS I[a]	CLASS II[a]	CLASS III
1,2	Not Permitted	Not Permitted	Permitted
3,4 (except Marine 4)	Not Permitted	Permitted[c]	Permitted
5,6,7,8, Marine 4	Permitted[b]	Permitted[c]	See Table R702.7(3)

a. Class I and II vapor retarders with vapor permeance greater than 1 perm when measured by ASTM E96 water method (Procedure B) shall be allowed on the interior side of any frame wall in all climate zones.
b. Use of a Class I interior vapor retarder in frame walls with a Class I vapor retarder on the exterior side shall require an approved design.
c. Where a Class II vapor retarder is used in combination with foam plastic insulating sheathing installed as continuous insulation on the exterior side of frame walls, the continuous insulation shall comply with Table R702.7(4) and the Class II vapor retarder shall have a vapor permeance greater than 1 perm when measured by ASTM E96 water method (Procedure B).

TABLE ~~R702.7.1~~ R702.7(3) Class III Vapor Retarders

Climate Zone	Class III Vapor Retarders Permitted For:[a,b]
7 ~~and 8~~	Continuous insulation with R-value ≥ 10 over 2 × 4 wall.
	Continuous insulation with R-value ≥ 15 over 2 × 6 wall.
8	Continuous insulation with R-value ≥ 12.5 over 2 × 4 wall.
	Continuous insulation with R-value ≥ 20 over 2 × 6 wall.

(no changes for Climate Zones Marine 4, 5 and 6)
a. Vented cladding shall include vinyl, polypropylene, horizontal aluminum siding, or brick veneer with a clear airspace as specified in Table R703.8.4(1), and other approved vented claddings.
b. The requirements in this table apply only to insulation used to control moisture in order to permit the use of Class III vapor retarders. The insulation materials used to satisfy this option also contribute to but do not supersede the thermal envelope requirements of Chapter 11.

R702.7(4) Continuous Insulation with Class II Vapor Retarder

Climate Zone	Class II Vapor Retarders Permitted for:[a]
3	Continuous insulation with R-value ≥ 2.
4, 5 and 6	Continuous insulation with R-value ≥ 3 over 2 × 4 wall.
	Continuous insulation with R-value ≥ 5 over 2 × 6 wall.
7	Continuous insulation with R-value ≥ 5 over 2 × 4 wall.
	Continuous insulation with R-value ≥ 7.5 over 2 × 6 wall.
8	Continuous insulation with R-value ≥ 7.5 over 2 × 4 wall.
	Continuous insulation with R-value ≥ 10 over 2 × 6 wall.

a. The requirements in this table apply only to insulation used to control moisture in order to permit the use of Class II vapor retarders. The insulation materials used to satisfy this option also contribute to but do not supersede the thermal envelope requirements of Chapter 11.

R702.7.1 Spray foam plastic insulation for moisture control with Class II and III vapor retarders. For purposes of compliance with Tables R702.7(3) and R702.7(4), spray foam with a maximum permeance of 1.5 perms at the installed thickness applied to the interior side of wood structural panels, fiberboard, insulating sheathing or gypsum shall be deemed to meet the continuous insulation moisture control requirement in accordance with one of the following conditions:

1. The spray foam R-value is equal to or greater than the specified continuous insulation R-value.
2. The combined R-value of the spray foam and continuous insulation is equal to or greater than the specified continuous insulation R-value.

(Deleted text in Section not shown for brevity and clarity.)

CHANGE SIGNIFICANCE: A wall must be designed to prevent accumulation of water within the wall assembly. This is accomplished by providing a water-resistant barrier behind exterior cladding with a means of draining to the exterior any water that enters the assembly. A vapor retarder is provided on the interior side of framed walls to resist passage of condensation into a wall assembly from the building interior. Water vapor tends to move from the warm side of an assembly to the cool side.

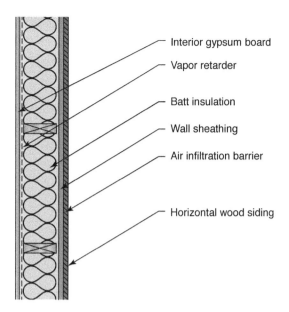

Vapor retarder on exterior wall.

The codes have historically required vapor retarders on the "warm in winter" side of exterior walls. Over time, research has shown that this is not needed in warmer climates. Therefore, the *International Residential Code* (IRC) requires a vapor retarder to be installed on the interior side of a building's thermal envelope in specific climate zones.

Three classifications of vapor retarders are referenced in the IRC: Class I, Class II and Class III.

- Class I vapor retarders are typically sheet polyethylene and nonperforated aluminum foil with a perm rating of more than 0.1.
- Class II retarders can include kraft-faced fiberglass batt insulation or paint that has a perm rating greater than 0.1 and no more than 1.0.
- Latex or enamel paint is considered a Class III vapor retarder with a perm rating between 1 and 10.

The IRC generally requires a Class I or Class II vapor retarder on the interior side of frame walls in Zones 5, 6, 7, 8 and Marine 4; although a Class III vapor retarder can be used when continuous insulation is used on the exterior side of the wall.

The 2021 IRC vapor retarder provisions are reorganized to clarify their use. New tables and text are intended to assist a designer in selecting appropriate vapor retarders for various climatic conditions and desired vapor retarder classes. Some vapor retarders do not fit neatly into categories defined by vapor permeability measurements alone. The prohibition of certain vapor retarder classes in warmer climate zones is based on the need to design assemblies with the ability to dry to the interior and not trap moisture within the wall that has come from outside the building.

However, if a vapor retarder has vapor permeability that increases with relative humidity (RH) to a Class III level, it will allow moisture movement through the exterior wall without accumulation and should not be prohibited. While there is adequate experience in colder climates in the United States and Canada to justify use of Class I interior vapor retarders, for example a 4-mil polyethylene, with foam plastic insulating sheathing on the exterior side of an assembly, a design for double vapor barrier assemblies must be submitted for approval. Due to concerns with low drying potential, walls using a double vapor barrier, in other words a Class I vapor retarder on both sides, must be accompanied by air sealing details and a drainage plane to minimize the potential for water leakage and moisture accumulation.

R703.2, R703.7.3
Water-Resistive Barriers

CHANGE TYPE: Modification

CHANGE SUMMARY: Language for water-resistive barriers is clarified with wet or dry climates specifically considered.

2021 CODE TEXT: **R703.2 Water-resistive barrier.** Not fewer than one layer of water-resistive barrier shall be applied over studs or sheathing of all exterior walls with flashing as indicated in Section R703.4, in such a manner as to provide a continuous water-resistive barrier behind the exterior wall veneer. The water-resistive barrier material shall be continuous to the top of walls and terminated at penetrations and building appendages in a manner to meet the requirements of the exterior wall envelope as described in Section R703.1.

Water-resistive barrier materials shall comply with one of the following:

1. No. 15 felt complying with ASTM D226, Type 1
2. ASTM E2556, Type I or II
3. ASTM E331 in accordance with Section R703.1.1
4. Other approved materials installed in accordance with the manufacturer's installation instructions.

No.15 asphalt felt and water-resistive barriers complying with ASTM E2556 shall be applied horizontally, with the upper layer lapped over the lower layer not less than 2 inches (51 mm), and where joints occur, shall be lapped not less than 6 inches (152 mm).

R703.7.3 Water-resistive barriers. Water-resistive barriers shall be installed as required in Section R703.2 and, where applied over wood-based sheathing, shall comply with Section R703.7.3.1 or Section R703.7.3.2.

House wrap installed as a continuous water barrier.

R703.7.3.1 Dry Climates. In dry (B) climate zones indicated in Figure N1101.7, water-resistive barriers shall comply with one of the following:

1. The water-resistive barrier shall be two layers of 10-minute Grade D paper or have a water resistance equal to or greater than two layers of a water-resistive barrier complying with ASTM E2556, Type I. The individual layers shall be installed independently such that each layer provides a separate continuous plane. Flashing installed in accordance with Section R703.4 and intended to drain to the water-resistive barrier, shall be directed between the layers.

2. The water-resistive barrier shall be 60-minute Grade D paper or have a water resistance equal to or greater than one layer of a water-resistive barrier complying with ASTM E2556, Type II. The water-resistive barrier shall be separated from the stucco by a layer of foam plastic insulating sheathing or other non-water-absorbing layer or a designed drainage space.

R703.7.3.2 Moist or marine climates. In the moist (A) or marine (C) climate zones indicated in Figure N1101.7, water-resistive barriers shall comply with one of the following:

1. In addition to complying with Section R703.7.3.1, a space or drainage material not less than 3/16 inch (5 mm) in depth shall be added to the exterior side of the water-resistive barrier.

2. In addition to complying with Section R703.7.3.1 Item 2, drainage on the exterior side of the water-resistive barrier shall have a drainage efficiency of not less than 90 percent, as measured in accordance with ASTM E2273 or Annex A2 of ASTM E2925.

(Deleted text not shown for brevity and clarity.)

CHANGE SIGNIFICANCE: Walls must be designed to prevent accumulation of water within wall assemblies. This is accomplished by providing a water-resistant barrier behind exterior cladding with a means of draining to the exterior any water that enters the assembly. A vapor retarder is provided on the interior side of exterior framed walls to resist passage of condensation into wall assemblies from the building interior.

The most common way to provide a drainage plane is to install weather-resistant materials in shingle fashion beneath a wall covering. A water-resistive barrier must provide a path for water flow if it migrates through a wall covering. Siding and veneers are typically not impervious to wind-driven rain. Water-resistive barriers in combination with flashing complete a weather protective system to move moisture out of a wall assembly. A water-resistive barrier must be continuous from the base of a building to the top of its walls.

There is a broad range of water-resistive barriers available in addition to traditional No. 15 asphalt felt. Sheet-type water-resistive barriers complying with ASTM D226, ASTM E331 or ASTM E2556 may be used. Nonsheet-style materials that provide equivalent protection may also be used. Section R703.2 now lists materials which may be used and testing standards that alternative materials must meet to be used as water-resistive barriers.

Section R703.7.3, water-resistive barriers for stucco, is reorganized into subsections that focus on different methods of complying with requirements based on wet or dry climatic conditions. It is critical to determine the local climate and tendency of moisture to infiltrate walls. These considerations were not accounted for in earlier code editions when stucco was assumed to be typically applied in dry, warm climates. Today, stucco is used throughout the United States and climate consideration is needed to avoid increased risk of moisture problems in climates that are moist or frequently rainy. This revision helps resolve performance issues due to moisture when stucco is placed over wood-based sheathing and does not affect construction methods in dry climates such as the southwestern region of the United States where stucco has performed well.

Table R703.8.4(1)
Veneer Attachment

CHANGE TYPE: Modification

CHANGE SUMMARY: Larger air gaps are allowed behind veneer to accommodate thicker continuous insulation.

2021 CODE TEXT:

TABLE R703.8.4(1) Tie Attachment and Airspace Requirements

Backing and Tie	Minimum Tie	Minimum Tie Fastener[a]	Airspace[b]	
Wood stud backing with corrugated sheet metal	22 U.S. gage (0.0299 in.) × 7/8 in. wide	8d common nail[bc] (2½ in. × 0.131in.)	Nominal 1 in. between sheathing and veneer	
Wood stud backing with adjustable metal strand wire	W1.7 (No. 9 U.S. gage; 0.148 in. dia.) with hook embedded in mortar joint[d]	8d common nail[bc] (2½ in. × 0.131in.)	Minimum nominal 1in. between sheathing and veneer	Maximum 4½ 4⅝ in. between backing and veneer
Wood stud backing with adjustable metal strand wire	W2.8 (0.187 in. dia.) with hook embedded in mortar joint[e,f]	8d common nail[c] (2½ in. × 0.131in.)	Greater than 4⅝ in. between backing and veneer	Maximum 6⅝ in. between backing and veneer
Cold-formed steel stud backing with adjustable metal strand wire	W1.7 (No. 9 U.S. gage; 0.148 in. dia.) with hook embedded in mortar joint[d]	No. 10 screw extending through the steel framing a minimum of three exposed threads	Minimum nominal 1 in. between sheathing and veneer	Maximum 4½ 4⅝ in. between backing and veneer
Cold-formed steel stud backing with adjustable metal strand wire	W2.8 (0.187 in. dia.) with hook embedded in mortar joint[e,f]	No. 10 screw extending through the steel framing a minimum of three exposed threads	Greater than 4⅝ in. between backing and veneer	Maximum 6⅝ in.between backing and veneer

b.a. All fasteners shall have rust-inhibitive coating suitable for the installation in which they are being used or be manufactured from material not susceptible to corrosion.

c.b. An airspace that provides drainage shall be permitted to contain mortar from construction.

a.c. In Seismic Design Category D_0, D_1 or D_2, the minimum tie fastener shall be an 8d ring-shank nail (2½ in. × 0.131 in.) or a No. 10 screw extending through the steel framing a minimum of three exposed threads.

d. Adjustable tie pintles shall include not fewer than 1 pintle leg of wire size W2.8 (MW18) with a maximum offset of 1-1/4 in.

e. Adjustable tie pintles shall include not fewer than 2 pintle legs with a maximum offset of 1¼ in. Distance between inside face of brick and end of pintle shall be a maximum of 2 in.

f. Adjustable tie backing attachment components shall consist of one of the following: eyes with minimum wire W2.8 (MW18), barrel with minimum ¼ in. outside dia., or plate with minimum thickness of 0.074 in. and minimum width of 1¼ in.

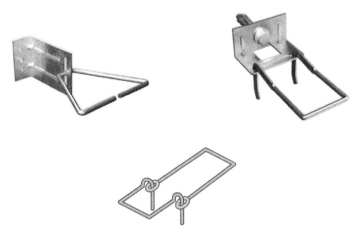

Veneer anchor examples.

CHANGE SIGNIFICANCE: IRC tie and airspace provisions now match *International Building Code* (IBC) requirements through reference to the anchored masonry veneer provisions of TMS 402, *Building Code Requirements for Masonry Structures*.

Energy conservation techniques using exterior foam sheathing require larger airspaces constructed between anchored masonry veneer and its backing in order to accommodate thicker continuous insulation used in colder climate zones. Masonry veneer with airspaces up to a maximum of 4$\frac{5}{8}$ inches may be constructed using traditional tie configurations. Airspaces up to 6$\frac{5}{8}$ inches must be constructed using stiffer tie configurations. These ties use thicker pintles with double legs or an eye, barrel or plate arrangement for greater strength.

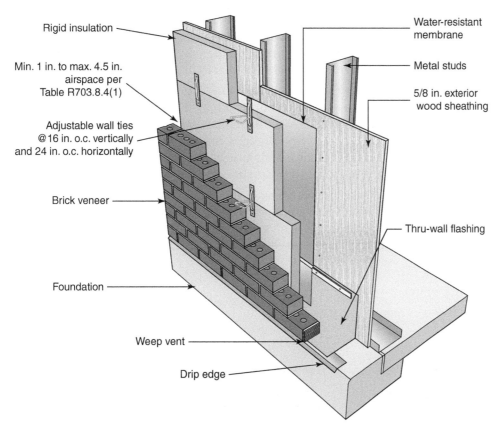

Veneer attachment to wood-frame construction.

R703.11.2
Vinyl Siding Installation Over Foam Plastic Sheathing

CHANGE TYPE: Modification

CHANGE SUMMARY: Wind pressure ratings for vinyl siding are decreased.

2021 CODE TEXT: **R703.11.2 Installation over foam plastic sheathing.** Where vinyl siding or insulated vinyl siding is installed over foam plastic sheathing, the vinyl siding shall comply with Section R703.11 and shall have a <u>wind load</u> design ~~wind~~ pressure ~~resistance~~ rating in accordance with Table R703.11.2.

Exceptions:

1. Where the foam plastic sheathing is applied directly over wood structural panels, fiberboard, gypsum sheathing or other approved backing capable of independently resisting the design wind pressure, the vinyl siding shall be installed in accordance with Sections R703.3.3 and R703.11.1.

2. Where the vinyl siding manufacturer's product specifications provide an approved <u>wind load</u> design ~~wind~~ pressure rating for installation over foam plastic sheathing, use of this <u>wind load</u> design ~~wind~~ pressure rating shall be permitted and the siding shall be installed in accordance with the manufacturer's installation instructions.

3. Where the foam plastic sheathing and its attachment have a design wind pressure resistance complying with Sections R316.8 and R301.2.1, the vinyl siding shall be installed in accordance with Sections R703.3.3 and R703.11.1.

Vinyl siding over foam plastic sheathing.

TABLE R703.11.2 ~~Adjusted~~ Required Minimum Wind Load Design ~~Wind~~ Pressure ~~Requirement~~ Rating for Vinyl Siding Installed over Foam Plastic Sheathing Alone

Ultimate Design Wind Speed (mph)	Adjusted Minimum Design Wind Pressure (ASD) (psf)[a,b]					
	Case 1: With interior gypsum wallboard[c]			Case 2: Without interior gypsum wallboard[c]		
	Exposure			Exposure		
	B	C	D	B	C	D
≤95	-30.0	-33.2	-39.4	-33.9	-47.4	-56.2
100	-30.0	-36.8	-43.6	-37.2	-52.5	-62.2
105	-30.0	-40.5	-48.1	-41.4	-57.9	-68.6
110	~~-44.0~~ -31.8	~~-61.6~~ -44.5	~~-73.1~~ -52.8	~~-62.9~~ -45.4	~~-88.1~~ -63.5	~~-104.4~~ -75.3
115	~~-49.2~~ -35.5	~~-68.9~~ -49.7	~~-81.7~~ -59.0	~~-70.3~~ -50.7	~~-98.4~~ -71.0	~~-116.7~~ -84.2
120	~~-51.8~~ -37.4	~~-72.5~~ -52.4	~~-86.0~~ -62.1	~~-74.0~~ -53.4	~~-103.6~~ -74.8	~~-122.8~~ -88.6
130	~~-62.2~~ -44.9	~~-87.0~~ -62.8	~~-103.2~~ -74.5	~~-88.8~~ -64.1	~~124.3~~ -89.7	~~-147.4~~ -106
> 130	See Footnote d ~~Not Allowed~~[d]					

For SI: 1 inch = 25.4 mm, 1 foot = 304.8 mm, 1 square foot = 0.0929 m², 1 mile per hour = 0.447 m/s, 1 pound per square foot = 0.0479 kPa.

a. Linear interpolation is permitted.
b. The table values are based on a maximum 30-foot mean roof height, and effective wind area of 10 square feet Wall Zone 5 (corner), and the ASD design <u>component and cladding</u> wind pressure from Table R301.2.1(1), adjusted for exposure in accordance with Table R301.2.1(2), multiplied by the following adjustment factors: ~~2.6~~ 1.87 (Case 1) and ~~3.7~~ 2.67 (Case 2) ~~for wind speeds less than 130 mph and 3.7 (Case 2) for wind speeds greater than 130 mph~~.
c. Gypsum wallboard, gypsum panel product or equivalent.
d. For the indicated wind speed condition, and where foam sheathing is the only sheathing on the exterior of frame walls with vinyl siding, ~~is not allowed unless the vinyl siding complies with an adjusted minimum design wind pressure requirement as determined in accordance with Note b and~~ the wall assembly ~~is~~ shall be capable of resisting an impact without puncture at least equivalent to that of a wood frame wall with minimum $^7/_{16}$-inch OSB sheathing as tested in accordance with ASTM E1886. The vinyl siding shall comply with an adjusted design wind pressure requirement in accordance with Note b, using an adjustment factor of 2.67.

CHANGE SIGNIFICANCE: Table R703.11.2 is updated so adjusted vinyl siding design wind pressure ratings are consistent with requirements in ASTM D3679, *Standard Specification for Rigid Poly (Vinyl Chloride) (PVC) Siding*. The ASTM D3679 pressure equalization factor for determining design wind pressure ratings of vinyl siding was recently revised. Because the pressure equalization factor in ASTM D3679 is now more conservative (changed from 0.36 to 0.5), adjustments for vinyl siding applications over foam sheathing are adjusted downward by multiplying existing tabulated values by 0.36/0.5 = 0.72. Consistency with ASTM D3679 is also provided by using the term wind load design pressure rating.

R704
Soffits

Fiber-cement soffit.

CHANGE TYPE: Addition

CHANGE SUMMARY: Requirements for soffit material and installation are expanded.

2021 CODE TEXT: R703.3.1 Soffit installation. Soffits shall comply with Section R704.~~R703.3.1.1, Section R703.3.1.2 or the manufacturer's installation instructions~~.

SECTION R704 SOFFITS

R704.1 General wind limitations. <u>Where the design wind pressure is 30 pounds per square foot (1.44 kPa) or less, soffits shall comply with Section R704.2. Where the design wind pressure exceeds 30 pounds per square foot (1.44 kPa), soffits shall comply with Section R704.3. The design wind pressure on soffits shall be determined using the component and cladding loads specified in Table R301.2.1(1) for walls using an effective wind area of 10 square feet and adjusted for height and exposure in accordance with Table R301.2.1(2).</u>

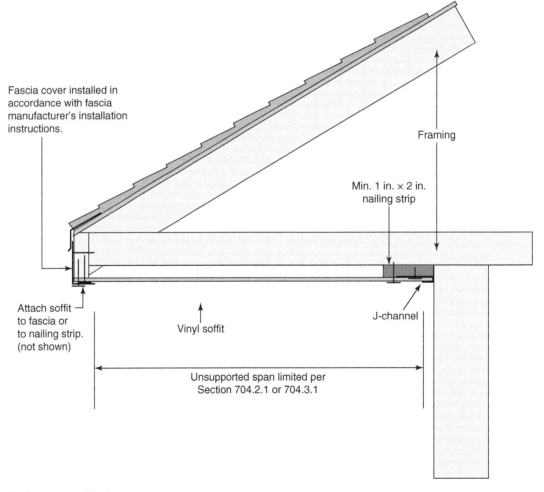

Soffit support: Single span.

R704.2 Soffit installation where the design wind pressure is 30 psf or less. Where the design wind pressure is 30 pounds per square foot (1.44 kPa) or less, soffit installation shall comply with Section R704.2.1, Section R704.2.2, Section R704.2.3, or Section R704.2.4. Soffit materials not addressed in Sections R704.2.1 through R704.2.4 shall be in accordance with the manufacturer's installation instructions.

R704.2.1 Vinyl soffit panels. Vinyl soffit panels shall be installed using fasteners specified by the manufacturer and shall be fastened at both ends to a supporting component such as a nailing strip, fascia or subfascial component in accordance with Figure R704.2.1(1). Where the unsupported span of soffit panels is greater than 16 inches, intermediate nailing strips shall be provided in accordance with Figure R704.2.1(2). Vinyl soffit panels shall be installed in accordance with the manufacturer's installation instructions. Fascia covers shall be installed in accordance with the manufacturer's installation instructions.

R704.2.2 Fiber-cement soffit panels Fiber-cement soffit panels shall be a minimum of ¼-inch in thickness and shall comply with the requirements of ASTM C1186, Type A, minimum Grade II or ISO 8336, Category A, minimum Class 2. Panel joints shall occur over framing or over wood structural panel sheathing. Soffit panels shall be installed with spans and fasteners in accordance with the manufacturer's installation instructions.

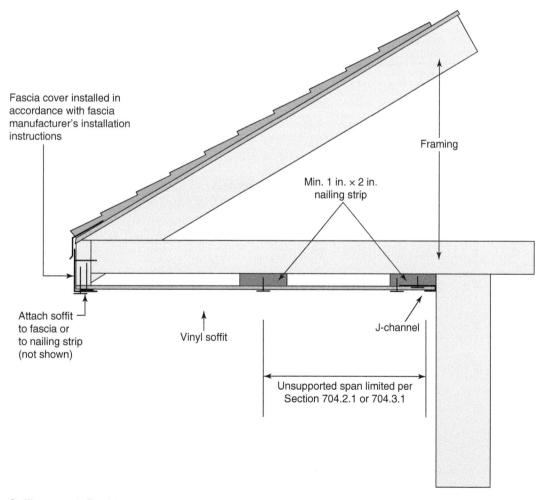

Soffit support: Double span.

R704.2.3 Hardboard soffit panels. Hardboard soffit panels shall be not less than $7/_{16}$-inch in thickness and shall be fastened to framing or nailing strips with 2½" × 0.113" (64 mm × 2.9 mm) siding nails spaced not more than 6 inches (152 mm) on center at panel edges and 12 inches (305 mm) on center at intermediate supports.

R703.3.1.1 R704.2.4 Wood structural panel soffit. The minimum nominal thickness for wood structural panel soffits shall be $3/_{8}$-inch (9.5 mm) and shall be fastened to framing or nailing strips with 2-inch by 0.099-inch (51 mm × 2.5 mm) nails. Fasteners shall be spaced not less than 6 inches (152 mm) on center at panel edges and 12 inches (305 mm) on center at intermediate supports.

R704.3 Soffit installation where the design wind pressure exceeds 30 psf. Where the design wind pressure is greater than 30 psf, soffit installation shall comply with Section R704.3.1, Section R704.3.2, Section R704.3.3, or Section R704.3.4. Soffit materials not addressed in Sections R704.3.1 through R704.3.4 shall be in accordance with the manufacturer's installation instructions.

R704.3.1 Vinyl soffit panels. Vinyl soffit panels and their attachments shall be capable of resisting wind loads specified in Table R301.2.1(1) for walls using an effective wind area of 10 square feet (0.929 m^2) and adjusted for height and exposure in accordance with Table R301.2.1(2). Vinyl soffit panels shall be installed using fasteners specified by the manufacturer and shall be fastened at both ends to a supporting component such as a nailing strip, fascia or subfascia component in accordance with Figure R704.2.1(1). Where the unsupported span of soffit panels is greater than 12 inches (305 mm), intermediate nailing strips shall be provided in accordance with Figure R704.2.1(2). Vinyl soffit panels shall be installed in accordance with the manufacturer's installation instructions. Fascia covers shall be installed in accordance with the manufacturer's installation instructions.

R704.3.2 Fiber-cement soffit panels. Fiber-cement soffit panels shall comply with Section R704.2.2 and shall be capable of resisting wind loads specified in Table R301.2.1(1) for walls using an effective wind area of 10 square feet (0.929 m^2) and adjusted for height and exposure in accordance with Table R301.2.1(2).

R704.3.3 Hardboard soffit panels. Hardboard soffit panels shall comply with the manufacturer's installation instructions and shall be capable of resisting wind loads specified in Table R301.2.1(1) for walls using an effective wind area of 10 square feet (0.929 m^2) and adjusted for height and exposure in accordance with Table R301.2.1(2).

R704.3.4 Wood structural panel soffit. Wood structural panel soffits shall be capable of resisting wind loads specified in Table R301.2.1(1) for walls using an effective wind area of 10 square feet (0.929 m^2) and adjusted for height and exposure in accordance with Table R301.2.1(2). Alternatively, wood structural panel soffits shall be installed in accordance with Table R704.3.4.

TABLE R704.3.4 Prescriptive Alternative for Wood Structural Panel Soffit[b,c,d,e]

Maximum Design Pressure (+ or - psf)	Minimum Panel Span Rating	Minimum Panel Performance Category	Nail Type and Size	Fastener[a] Spacing Along Edges and Intermediate Supports	
				Galvanized Steel	Stainless Steel
30	24/0	3/8	6d box (2 × 0.099 × 0.266 head diameter)	6[f]	4
40	24/0	3/8	6d box (2 × 0.099 × 0.266 head diameter)	6	4
50	24/0	3/8	6d box (2 × 0.099 × 0.266 head diameter)	4	4
			8d common (2½ × 0.131 × 0.281 head diameter)	6	6
60	24/0	3/8	6d box (2 × 0.099 × 0.266 head diameter)	4	3
			8d common (2½ × 0.131 × 0.281 head diameter)	6	4
70	24/16	7/16	8d common (2½ × 0.131 × 0.281 head diameter)	4	4
			10d box (3 × 0.128 × 0.312 head diameter)	6	4
80	24/16	7/16	8d common (2½ × 0.131 × 0.281 head diameter)	4	4
			10d box (3 × 0.128 × 0.312 head diameter)	6	4
90	32/16	15/32	8d common (2½ × 0.131 × 0.281 head diameter)	4	3
			10d box (3 × 0.128 × 0.312 head diameter)	6	4

For SI: 1 inch = 25.4 mm, 1 pound per square foot = 0.0479 kPa.

a. Fasteners shall comply with Sections R703.3.2 and R703.3.3.
b. Maximum spacing of soffit framing members shall not exceed 24 inches.
c. Wood structural panels shall be of an exterior exposure grade.
d. Wood structural panels shall be installed with strength axis perpendicular to supports with not fewer than two continuous spans.
e. Wood structural panels shall be attached to soffit framing members with specific gravity of at least 0.42. Framing members shall be minimum 2×3 nominal with the larger dimension in the cross section aligning with the length of fasteners to provide sufficient embedment depths.
f. Spacing at intermediate supports shall be not greater than 12 inches on center.

CHANGE SIGNIFICANCE: Specific material requirements are added for soffits to clarify their wind performance when using IRC installation provisions for common manufactured soffit types. New Section R704 also provides a prescriptive alternative for wood structural panel soffits that complies with design wind pressures specified in the IRC and ASCE 7, *Minimum Design Loads and Associated Criteria for Buildings and Other Structures*. This section refines and further clarifies provisions adopted into the 2018 IRC and adds new provisions to address soffit installation in high wind regions. Provisions move to new Section R704 to better distinguish soffit requirements from exterior wall covering provisions in Section R703.

Observations during damage assessments of Hurricane Harvey in Texas and Hurricane Irma in Florida revealed frequent soffit damage in both new and existing construction. In Florida, vinyl and metal soffits were damaged, but vinyl soffit panels were the most common product observed, particularly in the Florida Keys where vinyl soffit damage was widespread. In Texas, many soffits lacked adequate wind resistance due to use of improper materials, a lack of fasteners, or inadequate framing leading to water intrusion inside the building envelope. Based on estimated wind speeds at the sites visited, failure occurred to soffit components at wind speeds well below design wind speeds for the area. Loss of soffit vents can allow hurricane winds to drive large amounts of water through resulting openings and soak insulation, which can lead to mold growth and, in some cases, ceiling collapse.

In many cases, inadequate support and attachment at soffit panel ends led to failure. There were cases where soffits appeared to have been fastened by a single nailing strip across the midpoint of the framing above. New Section R704.2.1 clarifies that vinyl soffit panels are required to be fastened at each end and an unsupported span cannot exceed 16 inches unless permitted by the manufacturer's product approval.

Provisions in this section are applied based on component and cladding design wind pressures from Table R301.2.1(1) and Table R301.2.1(2) with an assumed 30 psf design wind pressure. For most of the country, design wind pressures will be 30 psf or less, even in Exposure Category D areas, requiring connections similar to those in the 2018 IRC. For higher wind regions, enhanced soffit construction will be required where the design wind pressure exceeds 30 psf.

R802
Wood Roof Framing

CHANGE TYPE: Modification

CHANGE SUMMARY: Revised provisions clarify ridge beam and ceiling joist requirements.

2021 CODE TEXT: R802.3 Ridge. A ridge board used to connect opposing rafters shall be not less than 1 inch (25 mm) nominal thickness and not less in depth than the cut end of the rafter. Where ceiling joist or rafter ties do not provide continuous ties across the structure <u>as required by Section R802.5.2, the</u> ~~a~~ ridge <u>shall be supported by a wall or ridge beam designed in accordance with accepted engineering practice</u> ~~shall be provided~~ and supported on each end by a wall or <u>column</u> ~~girder~~.

R802.5 Ceiling joists. Ceiling joists shall be continuous across the structure or securely joined where they meet over interior partitions in accordance with ~~Table R802.5.2~~ <u>Section R802.5.2.1. Ceiling joists shall be fastened to the top plate in accordance with Table R602.3(1).</u>

R802.5.2 Ceiling joist and rafter connections. Where ceiling joists run parallel to rafters, <u>and are located</u> ~~they shall be connected to rafters at the top wall plate in accordance with Table R802.5.2. Where ceiling joists are not connected to the rafters at the top wall plate, they shall be installed~~ in the bottom third of the rafter height, <u>they shall be installed</u> in accordance with Figure R802.4.5 and <u>fastened to rafters in accordance with</u> Table R802.5.2(1). Where the ceiling joists are installed above the bottom third of the rafter height, the ridge shall be designed as a beam <u>in accordance with R802.3</u>. Where ceiling joists do not run parallel to rafters, ~~the ceiling joists shall be connected to top plates in accordance with Table R602.3(1). Each~~ rafter<u>s</u> shall be tied across the structure with a rafter tie <u>in accordance with R802.5.2.2</u>, or the ridge shall be designed

Wood-framed roof line.

as a beam in accordance with R802.3. ~~or a 2-inch by 4-inch (51 mm × 102 mm) kicker connected to the ceiling diaphragm with nails equivalent in capacity to Table R802.5.2.~~

R802.5.2.1 Ceiling joists lapped. Ends of ceiling joists shall be lapped not less than 3 inches (76 mm) or butted over bearing partitions or beams and toenailed to the bearing member. Where ceiling joists are used to provide <u>the continuous tie across the building</u> ~~resistance to rafter thrust~~, lapped joists shall be nailed together in accordance with Table R802.5.2(1) and butted joists shall be tied together <u>with a connection of equivalent capacity</u> ~~in a manner to resist such thrust~~. <u>Laps in joists</u> ~~Joists~~ that do not ~~resist thrust~~ <u>provide the continuous tie across the building</u> shall be permitted to be nailed in accordance with Table R602.3(1). ~~Wood structural panel roof sheathing, in accordance with Table R503.2.1.1(1), shall not cantilever more than 9 inches (229 mm) beyond the gable endwall unless supported by gable overhang framing.~~

R802.5.2.2 Rafter ties. Wood rafter ties shall be not less than 2 inches by 4 inches (51 mm × 102 mm) installed in accordance with Table R802.5.2(1) ~~at each rafter~~ <u>a maximum of 24 inches (610 mm) on center.</u> Other approved rafter tie methods shall be permitted.

R802.5.2.3 Blocking. Blocking shall be not less than utility grade lumber.

CHANGE SIGNIFICANCE: Requirements for connection of rafters and ceiling joists are clarified. Section R802.5 establishes the concept of a continuous tie across lower portions of rafters, using either ceiling joists or rafter ties, so rafters under roof loads do not push out exterior bearing walls—an action referred to as rafter thrust. Rafters and ceiling joists form a triangle, which is a stable shape as long as member ends are connected together. When the lower ends of rafters are not connected to rafter ties or ceiling joists to resist thrust, the only way to prevent thrust is to support them at the peak by a ridge beam.

Ceiling joists are installed in the lower portion of an attic and fastened in a specific manner to create a rafter tie. However, ceiling joists may be installed higher in an attic or perpendicular to rafters. Occasionally, there may not be any ceiling joists, such as in a cathedral ceiling. When ceiling joists are not located at the top of a bearing wall, ties are required to meet the provisions of Section R802.5.2 to ensure that rafters do not push (thrust) walls outward. In these cases, either a tie can be provided, or a ridge beam should be designed in accordance with accepted engineering practice.

CHANGE TYPE: Modification

CHANGE SUMMARY: The heel joint connection table is updated for roof spans of 24 and 36 feet and a 19.2-inch rafter spacing.

2021 CODE TEXT:

TABLE R802.5.2(1) Rafter/Ceiling Joist Heel Joint Connections[g]

		GROUND SNOW LOAD (psf)										
		20[e]			30			50			70		
		Roof span (feet)										
RAFTER SLOPE	RAFTER SPACING (inches)	12	24	36	12	24	36	12	24	36	12	24	36
		Required number of 16d common nails per heel joint splice[a,b,c,d,f]										
3:12	12	3	5	8	3	6	9	5	9	13	6	12	17
	16	4	7	10	4	8	12	6	12	17	8	15	23
	19.2	4	8	12	5	10	14	7	14	21	9	18	27
	24	5	10	15	6	12	18	9	17	26	12	23	34
4:12	12	3	4	6	3	5	7	4	7	10	5	9	13
	16	3	5	8	3	6	9	5	9	13	6	12	17
	19.2	3	6	9	4	7	11	6	11	16	7	14	21
	24	4	8	11	5	9	13	7	13	19	9	17	26
5:12	12	3	3	5	3	4	6	3	6	8	4	7	11
	16	3	4	6	3	5	7	4	7	11	5	9	14
	19.2	3	5	7	3	6	9	5	9	13	6	11	17
	24	3	6	9	4	7	11	6	11	16	7	14	21
7:12	12	3	3	4	3	3	4	3	4	6	3	5	8
	16	3	3	5	3	4	5	3	5	8	4	7	10
	19.2	3	4	5	3	4	6	3	6	9	4	8	12
	24	3	5	7	3	5	8	4	8	11	5	10	15
9:12	12	3	3	3	3	3	3	3	3	5	3	4	6
	16	3	3	4	3	3	4	3	4	6	3	5	8
	19.2	3	3	4	3	4	5	3	5	7	3	6	9
	24	3	4	5	3	4	6	3	6	9	4	8	12
12:12	12	3	3	3	3	3	3	3	3	4	3	3	5
	16	3	3	3	3	3	3	3	3	5	3	4	6
	19.2	3	3	3	3	3	4	3	4	6	3	5	7
	24	3	3	4	3	3	5	3	5	7	3	6	9

For SI: 1 inch = 25.4 mm, 1 foot = 304.8 mm, 1 pound per square foot = 0.0479 kPa.

a. 10d common (3" × 0.148") nails shall be permitted to be substituted for 16d common (3-1/2" × 0.162") nails where the required number of nails is taken as 1.2 times the required number of 16d common nails, rounded up to the next full nail. 40d box nails shall be permitted to be substituted for 16d common nails.
b. Nailing requirements shall be permitted to be reduced 25 percent if nails are clinched.
b.c. Heel joint connections are not required where the ridge is supported by a load-bearing wall, header or ridge beam.
c.d. Where intermediate support of the rafter is provided by vertical struts or purlins to a load-bearing wall, the tabulated heel joint connection requirements shall be permitted to be reduced proportionally to the reduction in span.
d.e. Equivalent nailing patterns are required for ceiling joist to ceiling joist lap splices.
e.f. Applies to roof live load of 20 psf or less.
f.g. Tabulated heel joint connection requirements assume that ceiling joists or rafter ties are located at the bottom of the attic space. Where ceiling joists or rafter ties are located higher in the attic, heel joint connection requirements shall be increased by the adjustment factors in Table R802.5.2(2).
g. Tabulated requirements are based on 10 psf roof dead load in combination with the specified roof snow load and roof live load.

TABLE R802.5.2(2) Heel Joint Connection Adjustment Factors

$H_C/H_R{}^{a,b}$	Heel Joint Connection Adjustment Factor
1/3	1.5
1/4	1.33
1/5	1.25
1/6	1.2
1/10 or less	1.11

a. H_C = Height of ceiling joists or rafter ties measured vertically ~~above~~ from the top of the rafter support walls to the bottom of the ceiling joists or rafter ties; H_R = Height of roof ridge measured vertically ~~above~~ from the top of the rafter support walls to the bottom of the roof ridge.

b. Where H_C/H_R exceeds 1/3, connections shall be designed in accordance with accepted engineering practice.

(Deleted text in table not shown for brevity and clarity.)

Heel joint connection.

CHANGE SIGNIFICANCE: Table R802.5.2(1) is updated to be consistent with the calculation basis of the American Wood Council's (AWC) *2018 Wood Frame Construction Manual* (WFCM) heel joint nailing requirements which are based on the AWC *2018 National Design Specification for Wood Construction* (NDS) provisions for nailed connections. A reduction in the required number of 16d common nails required in rafter tie connections by approximately 15 percent is due to changes based on a penetration factor and load duration. Previously, a 0.77 penetration factor (based on the 1991 and 1997 NDS) was used for 16d common nails with less than 12d penetration in the main member and a load duration factor of 1.25.

Revised nailing requirements are based on a 1.15 load duration factor for snow load cases, 1.25 load duration factor for roof live load cases, and an effective penetration factor equal to 1.0 per the current NDS where lateral design value calculations for nails are based on the actual penetration in the wood member. The ratio of nail design values for snow cases originally used to develop nailing requirements to the current nail design values for snow cases is $(Z \times 0.77 \times 1.25) / (Z \times 1.0 \times 1.15) = 0.84$ and explains the reduced number of nails now required in the table. Due to revised NDS nail design provisions, clinched nails are no longer recognized for this application. A 10d common nail option is added based on NDS lateral design value calculations.

Lastly, a sentence is added to footnote f, now Table R802.5.2(2), to clarify that rafter tie connections higher than the bottom third of the attic space ($H_C/H_R > 1/3$) must be engineered while the definitions of H_C and H_R are clarified to show how the variables should be measured.

R802.6
Rafter and Ceiling Joist Bearing

CHANGE TYPE: Modification

CHANGE SUMMARY: Text is added to clarify when a ridge board connection is sufficient for bearing.

2021 CODE TEXT: R802.6 Bearing. The ends of each rafter or ceiling joist shall have not less than $1^1/_2$ inches (38 mm) of bearing on wood or metal and not less than 3 inches (76 mm) on masonry or concrete. The bearing on masonry or concrete shall be direct, or a sill plate of 2-inch (51 mm) minimum nominal thickness shall be provided under the rafter or ceiling joist. The sill plate shall provide a minimum nominal bearing area of 48 square inches (30 865 mm^2). <u>Where the roof pitch is greater than or equal to 3 units vertical in 12 units horizontal (25-percent slope), and ceiling joists or rafter ties are connected to rafters to provide a continuous tension tie in accordance with Section R802.5.2, vertical bearing of the top of the rafter against the ridge board shall satisfy this bearing requirement.</u>

CHANGE SIGNIFICANCE: Changes to Section R802.6 clarify acceptable bearing area, or contact, for rafters, specifically at the top ends of rafters. Section R802.6 requires that the ends of each rafter or ceiling joist have at least 1½ inches of bearing on wood or metal. Bearing typically occurs when one component of an assembly rests on a horizontal surface, such as a top plate, beam or hanger of another element in an assembly to resist vertical loads. However, for a rafter system that has collar ties (or straps) at the top and a continuous tension tie at the bottom of the roof assembly provided by ceiling joists or rafter ties, the downward (gravity) force is transferred to the lower ends of the rafters, and a horizontal bearing force occurs at the top of the rafter toward the ridge board.

Rafters bearing on exterior side wall.

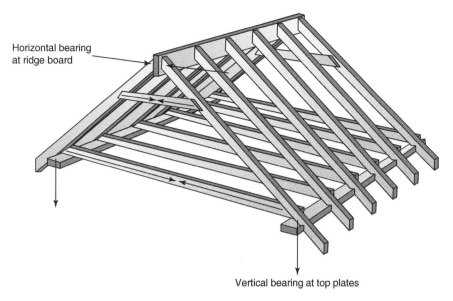

Forces in a stick-built roof.

When a roof slope is shallow, a structural ridge beam is required to manage horizontal and vertical forces per Section R802.4.4. In this case, the ridge beam provides vertical support for the top end of the rafters, and a connection with vertical capacity is required. An example is a joist hanger that provides a horizontal bearing surface for the end of the rafter.

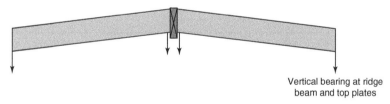

Forces in a low-slope roof.

Cathedral ceilings, while typically having a higher slope, normally do not have ceiling joists or collar ties. Since a horizontal tie is not provided at the bottom of the rafters, a load-bearing ridge beam is required per Section R802.5.2. A ridge beam provides vertical support (bearing) for the top end of the rafter.

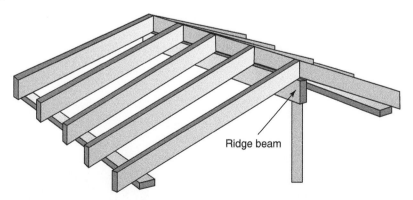

Cathedral ceilings require a ridge beam to resist vertical forces.

Table R804.3

CFS Roof Framing Fasteners

CHANGE TYPE: Modification

CHANGE SUMMARY: Connections for cold-formed steel (CFS) roof framing members are updated and clarified.

2021 CODE TEXT:

TABLE R804.3 Roof Framing Fastening Schedule[a,b]

Description of Building Elements			Number and Size of Fasteners[a]				Spacing of Fasteners
Roof sheathing (oriented strand board or plywood) to rafter			No. 8 screws				6" o.c. on edges and 12" o.c. at interior supports. 6" o.c. at gable end truss
Gypsum board to ceiling joists			No. 6 screws				12" o.c.
Gable end truss to endwall top track			No. 10 screws				12" o.c.
Rafter to ceiling joist and to ridge member			Minimum No. 10 screws, in accordance with Table R804.3.1.1(3)				Evenly spaced, not less than 1/2" from all edges.
Ceiling joist or roof truss to top track of bearing wall[b]			Ultimate Design Wind Speed (mph) and Exposure Category				Each ceiling joist or roof truss
	Ceiling Joist or Truss Spacing (in.)	Roof Span (ft)	130B 115C	<139B 120C	130C	<139C	
	16	24	3	3	4	5	
		28	3	3	4	5	
		32	3	4	5	6	
		36	4	4	5	6	
		40	4	4	6	7	
	24	24	4	5	6	7	
		28	4	5	6	8	
		32	4	6	7	8	
		36	4	6	8	9	
		40	6	6	8	10	

For SI: 1 inch = 25.4 mm, 1 foot = 304.8 mm, 1 pound per square foot = 0.0479 kPa, 1 mil = 0.0254 mm.

a. Screws are a minimum No. 10 unless noted otherwise.

b. Indicated number of screws shall be applied through the flanges of the truss or ceiling joist or through each leg of a 54-mil clip angle. See Section R804.3.8 for additional requirements to resist uplift forces.

(Deleted text not shown for clarity.)

TABLE R804.3.2.1(2) Ultimate Design Wind Speed to Equivalent Snow Load Conversion

Ultimate Wind Speed and Exposure		Equivalent Ground Snow Load (psf) Roof slope									
Exposure	Wind speed (mph)	3:12	4:12	5:12	6:12	7:12	8:12	9:12	10:12	11:12	12:12
B	115	20	20	20	20	~~30~~ 20	~~30~~ 20	~~30~~ 20	~~30~~ 20	~~30~~ 20	~~50~~ 20
	120	20	20	20	20	~~30~~ 20	~~30~~ 20	~~30~~ 20	~~30~~ 20	~~30~~ 20	~~50~~ 20
	130	20	20	20	20	~~30~~ 20	~~30~~ 20	~~30~~ 20	~~50~~ 20	~~50~~ 20	~~50~~ 20
	<140	20	20	20	20	~~30~~ 20	~~50~~ 20	~~50~~ 20	~~50~~ 30	~~50~~ 30	~~50~~ 30
C	115	20	20	20	20	~~30~~ 20	~~30~~ 20	~~30~~ 20	~~50~~ 20	~~50~~ 30	~~50~~ 30
	120	20	20	20	20	~~30~~ 20	~~30~~ 20	~~50~~ 20	~~50~~ 30	~~50~~ 30	50
	130	20	20	~~20~~ 30	30	30	~~50~~ 30	~~50~~ 30	~~50~~ 30	50	~~70~~ 50
	<140	30	30	~~30~~ 50	50	~~50~~ 30	~~50~~ 30	~~70~~ 50	~~70~~ 50	~~70~~ 50	— 50

For SI: 1 mile per hour = 0.447 m/s, 1 pound per square foot = 0.0479 kPa.

R804.3.2.1.2 Rake overhangs. Rake overhangs shall not exceed ~~12 inches (305 mm) measured horizontally.~~ the limitations provided for Option 1 or Option 2 in Figure R804.3.2.1.2. Outlookers at gable endwalls shall be installed in accordance with Figure R804.3.2.1.2. The required strength for uplift connectors required for Option 1 shall be determined in accordance with AISI S230 Table F3-4.

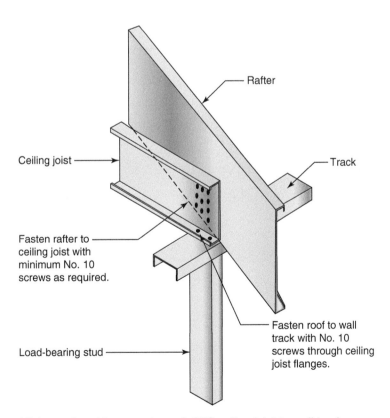

Minimum 3 to 10 screws in each CFS ceiling joist to wall track connection.

CHANGE SIGNIFICANCE: The cold-formed steel framing provisions of the *International Residential Code* (IRC) are updated to parallel requirements in the latest edition of AISI S230 - *Standard for Cold-Formed Steel Framing - Prescriptive Method for One- and Two-Family Dwellings* (AISI S230-18). The 2018 edition of AISI S230 has updated prescriptive provisions conforming to loading criteria of ASCE 7-16, *Minimum Design Loads and Associated Criteria for Buildings and Other Structures.* The cold-formed steel provisions affected by the latest edition of ASCE 7 are located in IRC Section 804 – Roof Framing. Wind pressure coefficients on roof surfaces have increased (see the significant change article for Table R301.2.1(1) for more detail). AISI S230-18 also reduced required bottom flange bracing spacing from 8 feet to 4 feet to minimize changes in the allowable roof span tables due to increased wind loading.

Table R804.3 adds fastening requirements for gypsum board to ceiling joist connections and rafter to ridge connections. Fastening requirements for wind resistance of the ceiling joist or truss to wall track are expanded with fastening increase with higher wind speeds or exposure category.

Table R804.3.2.1(2) updates the wind to ground snow load table for steeper pitched roofs. For example, the equivalent ground snow load is decreased from 30 psf to 20 psf for a roof pitch of 7:12 in a 115-mph region. Lastly, the rake overhang section is clarified by referencing limits in Figure R804.3.2.1.2 and minimum uplift connector capacity in AISI S230 Table F3-4.

R905.4.4.1 Metal Roof Shingle Wind Resistance

CHANGE TYPE: Addition

CHANGE SUMMARY: Requirements for metal shingle wind resistance are added to Section R905.4.

2021 CODE TEXT: R905.4.4.1 Wind Resistance of metal roof shingles. Metal roof shingles applied to a solid or closely fitted deck shall be tested in accordance with ASTM D3161, FM 4474, UL 580, or UL 1897. Metal roof shingles tested in accordance with ASTM D3161 shall meet the classification requirements of Table R905.4.4.1 for the appropriate maximum basic wind speed and the metal shingle packaging shall bear a label to indicate compliance with ASTM D3161 and the required classification in Table R905.2.4.1.

TABLE R905.4.4.1 Classification of Steep Slope Metal Roof Shingles Tested In Accordance With ASTM D3161

Maximum Ultimate Design Wind Speed, V_{ULT} From Figure R301.2(2) (mph)	Maximum Basic Wind Speed, V_{ASD} From Table R301.2.1.3 (mph)	ASTM D3161 Shingle Classification
110	85	A, D or F
116	90	A, D or F
129	100	A, D or F
142	110	F
155	120	F
168	130	F
181	140	F
194	150	F

For SI: 1 mile per hour = 1.609 kph.

Metal roof shingles.

Photo courtesy of Metal Roof Alliance

CHANGE SIGNIFICANCE: Under wind loads, metal roof shingles do not behave in the same manner as asphalt roof shingles defined in Section R905.2.4.1. Nonetheless, a classification similar to the asphalt shingle classification Table R905.2.4.1 is appropriate and was adapted for Section R905.4.4.1.

For metal roof components, wind uplift testing is currently addressed by multiple standards that determine compliance through uplift ratings. Metal roof shingle performance is not correctly represented by these current tests due to the air permeability inherent in the design of shingle units. A fan-induced test method was developed through ASTM as an alternative to metal roof panel uplift resistance testing.

ASTM D3161, *Standard Test Method for Wind Resistance of Steep Slope Roofing Products* (*Fan-Induced Method*), was originally created for asphalt shingles. The standard was expanded in 2013 to evaluate wind resistance of discontinuous, air-permeable, steep-slope roofing products with or without contribution from adhesives or mechanical interlocking to hold down the leading tab edge. ASTM D3161 is not limited to asphalt shingles and removes difficulties for metal shingle manufacturers previously required to test their products in accordance with UL 1897 or UL 580 via a nonair-permeable method not representative of the product. UL has provided metal shingle wind classifications for many years and currently has ASTM D3161-related listings in its Online Classification Directory. The ASTM test method is applicable to many other steep-slope roofing products and is used to evaluate the wind resistance of those products by testing and certification laboratories.

PART 4

Energy Conservation

Chapter 11

- **Chapter 11** Energy Efficiency

The IRC energy provisions are extracted from the residential provisions of the *International Energy Conservation Code* (IECC) and editorially revised to conform to the scope and application of the IRC. The section numbers appearing in parentheses after each IRC section number are the section numbers of the corresponding text in the IECC. The IECC Residential Provisions and Chapter 11 of the IRC provide for the effective use and conservation of energy in new residential buildings by regulating the building thermal envelope, mechanical systems, electrical systems, and service water heating systems. IRC Section N1101 establishes climate zones for geographical locations as the basis for determining thermal envelope requirements for conserving energy. The various elements of the building thermal envelope are covered in Section N1102 and include specific insulation, fenestration, and air-leakage requirements for improving energy efficiency. Section N1103 primarily is concerned with mechanical system controls, insulation and sealing of ductwork, equipment sizing, and mandatory mechanical ventilation systems. The insulation of mechanical and service hot water piping systems is also covered in the mechanical systems provisions of Section N1103. Energy-efficient lighting is covered in Section N1104. Alternative compliance paths using the total building performance and Energy Rating Index methods appear in Sections N1105 and N1106. The compliance path for tropical climate has moved to Section N1107. New provisions for additional energy efficiency package options are located in Section N1108. Sections N1109 through N1113 address the application of the energy provisions for work performed on existing buildings. ■

N1101.6
Definition of High-Efficacy Light Sources

N1101.7
Climate Zones

N1101.13
Compliance Options

N1101.13.5
Additional Energy Efficiency Requirements

N1101.14
Permanent Energy Certificate

N1102.1
Building Thermal Envelope

TABLES N1102.1.2 AND N1102.1.3
Insulation and Fenestration Requirements

N1102.2
Ceiling Insulation

N1102.2.4
Access Hatches and Doors

N1102.2.7
R-Value Reduction for Walls with Partial Structural Sheathing

N1102.2.7
Floor Insulation

N1102.2.8
Unconditioned Basement

N1102.4 AND TABLE N1102.4.1.1
Building Air Leakage and Testing

N1102.4.6
Air-Sealed Electrical Boxes

N1103.3
Duct Installation

N1103.3.5
Duct Testing

N1103.6
Mechanical Ventilation

N1104
Lighting Equipment

N1105 AND TABLE N1105.2
Total Building Performance Analysis

N1106 AND TABLE N1106.2
Energy Rating Index Analysis

N1108
Additional Efficiency Package Options

N1101.6
Definition of High-Efficacy Light Sources

CHANGE TYPE: Modification

CHANGE SUMMARY: The definition related to high-efficacy lighting now includes both lamps and luminaires and better reflects current technology and federal standards.

2021 CODE: HIGH-EFFICACY <s>LAMPS</s> <u>LIGHT SOURCES</u>. Compact fluorescent lamps, light-emitting diode (LED) lamps, T-8 or smaller diameter linear fluorescent lamps, <s>or</s> other lamps with an efficacy of not less than <s>the following:</s> <u>65 lumens per watt, or luminaires with an efficacy of not less than 45 lumens per watt.</u>

<s>1. 60 lumens per watt for lamps over 40 watts.</s>
<s>2. 50 lumens per watt for lamps over 15 watts to 40 watts.</s>
<s>3. 40 lumens per watt for lamps 15 watts or less.</s>

CHANGE SIGNIFICANCE: The definition of high-efficacy lamps first appeared in the 2009 edition of the code when the LED market was in its infancy. The definition has not changed to keep up with subsequent technology and market developments. As a result, the definition no longer represents the "high efficacy" share of the marketplace. New federal lighting standards in effect in 2020 will eliminate all bulbs on the market with efficiencies lower than 45 lumens per watt. Therefore, when the 2021 code is published, some of the bulbs currently defined by the IECC and Chapter 11 of the IRC as "high efficacy" will not be approved to sell. Given these market and federal standard changes, the definition has been updated to remain relevant and the term "lamp" has been changed to "light source" to be all inclusive. For example, luminaires are added to the definition. A luminaire, as defined by the electrical provisions and the NEC, is a complete lighting unit consisting of a light source, such as lamps (bulbs), that connects to power and distributes light. The requirements for the amount of high efficacy lighting appear in Section N1104.

High-efficacy LED lamps.

N1101.7
Climate Zones

CHANGE TYPE: Modification

CHANGE SUMMARY: The climate zone provisions have been comprehensively updated, including a new Climate Zone 0. Approximately 10 percent of U.S. counties have been assigned a new climate zone.

2021 CODE: N1101.7 (R301.1) Climate zones. Climate zones from Figure N1101.7 or Table N1101.7 shall be used for determining the applicable requirements in Sections N1101 through ~~N1111~~ N1113. Locations not indicated in Table N1101.7 shall be assigned a climate zone in accordance with Section N1101.7.2.

Note: Only portions of Section N1101.7 and Table N1101.7 are shown. For the complete text and table, refer to the 2021 IRC.

CHANGE SIGNIFICANCE: The climate zone provisions in Section N1107, including the climate zone tables and map have undergone a significant update. This update of the climate zones corresponds with ASHRAE Standard 169-2013, which is referenced in the 2018 *International Green Construction Code* (IgCC), ASHRAE 90.1, and ASHRAE 90.2. ICC has a licensing agreement with ASHRAE to include the climate zone map, definitions and tables for consistency with ASHRAE Standard 169-2013.

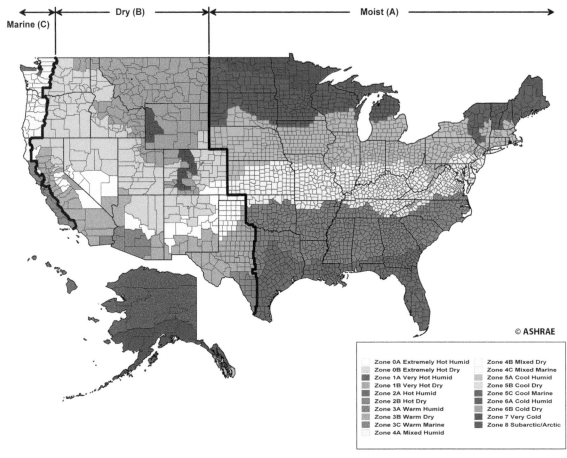

Climate Zones.

TABLE N1101.7 (R301.1) Climate Zones, Moisture Regimes, and Warm-Humid Designations by State, County and Territory[a]

US STATES

ALABAMA
3A 2A Coffee*
3A 2A Covington*
3A 2A Dale*
3A 2A Escambia*
3A 2A Geneva*
3A 2A Henry*
3A 2A Houston*

ALASKA
8 7 Bethel
7 Bristol Bay
7 8 Denali
8 7 Dillingham
7 6A Haines
7 6A Juneau
7 Kenai Peninsula
7 5C Ketchikan Gateway
7 6A Kodiak Island
7 5C Prince of Wales-
   Outer Ketchikan
7 5C Sitka
7 6A Skagway-Hoonah-Angoon
7 6A Wrangell-Petersburg

COLORADO
5B 4B Bent
5B Broomfield
6B 5B Custer
5B 4B Prowers

FLORIDA
2A 1A Palm Beach*

GEORGIA
4A 3A Banks
3A 2A Calhoun*

4A 3A Catoosa
4A 3A Chattooga
3A 2A Coffee*
4A 3A Dade
4A 3A Dawson
3A 2A Dougherty*
3A 2A Early*
4A 3A Fannin
4A 3A Floyd
4A 3A Franklin
4A 3A Gilmer
4A 3A Gordon
4A 3A Habersham
4A 3A Hall
4A 3A Lumpkin
4A 3A Murray
4A 3A Pickens
4A 3A Rabun
4A 3A Stephens
3A 2A Tift*
4A 3A Towns
4A 3A Union
4A 3A Walker
4A 3A White
4A 3A Whitfield
3A 2A Worth*

ILLINOIS
5A 4A Calhoun
5A 4A Clark
5A 4A Coles
5A 4A Cumberland
5A 4A Greene
5A 4A Jersey

INDIANA
5A 4A Bartholomew
5A 4A Clay
5A 4A Decatur

5A 4A Fayette
5A 4A Franklin
5A 4A Hendricks
5A 4A Johnson
5A 4A Marion
5A 4A Morgan
5A 4A Owen
5A 4A Putnam
5A 4A Rush
5A 4A Shelby
5A 4A Union
5A 4A Vigo

IOWA
6A 5A Allamakee
6A 5A Bremer
6A 5A Buchanan
6A 5A Buena Vista
6A 5A Butler
6A 5A Calhoun
6A 5A Cherokee
6A 5A Chickasaw
6A 5A Clayton
6A 5A Delaware
6A 5A Fayette
6A 5A Floyd
6A 5A Franklin
6A 5A Grundy
6A 5A Hamilton
6A 5A Hardin
6A 5A Howard
6A 5A Humboldt
6A 5A Ida
6A 5A Plymouth
6A 5A Pocahontas
6A 5A Sac
6A 5A Webster
6A 5A Winneshiek
6A 5A Wright

a. Key: A – Moist, B – Dry, C – Marine.   Absence of moisture designation indicates moisture regime is irrelevant. Asterisk (*) indicates a warm-humid location.

Two updates that are relevant to most code users include:

- Updating the climate zone designations in Table N1107 for approximately 10 percent of the counties in the United States. In general, the counties with changes were assigned a warmer climate zone.
- Updating Figure N1107, the climate zone map, to reflect the changes in the county's climate zone designations.

Notably, the number of climate zones has increased to include Climate Zone 0, representing the hottest of climates. Climate Zone 0 is a subset of the previous Climate Zone 1. Whereas the previous Climate Zone 1 was all locations with more than 9,000 Cooling Degree Days, Climate Zone 1 now "tops out" at 10,800 Cooling Degree Days, and Climate Zone 0 is for those locations with more than 10,800 Cooling Degree Days. Cities in Climate Zone 0 include extremely hot locations such as Mumbai (Bombay), Jakarta and Abu Dhabi. There are no cities in the United States in Climate Zone 0; Miami and the islands of Hawaii are in Climate Zone 1. Several code changes introduced specific provisions for Climate Zone 0. Because Climate Zone 0 is a subset of the former Climate Zone 1, references throughout the IECC (Chapter 11 of the IRC) to Climate Zone 1 have been edited to Climate Zone 0-1.

Code users needing to calculate the moisture zone (marine, dry or humid) will find Section N1107.3 entirely revised. Calculations are in three categories, (a) when monthly average temperature and precipitation data is available, (b) when annual average temperature, degree days and precipitation is available, and (c) when only degree day information is available.

N1101.13
Compliance Options

CHANGE TYPE: Modification

CHANGE SUMMARY: The compliance path options have been clarified and the Prescriptive and Mandatory labels in the section titles have been removed.

2021 CODE: N1101.13 (R401.2) ~~Compliance~~ **Application.** ~~Projects~~ Residential buildings shall comply with Section N1101.13.5 and Section N1101.13.1, N1101.13.2, N1101.13.3, or N1101.13.4. ~~one of the following:~~

1. ~~Sections N1101.14 through N1104.~~
2. ~~Section N1105 and the provisions of Sections N1101.14 through N1104 indicated as "Mandatory."~~
3. ~~The energy rating index (ERI) approach in Section N1106.~~

Exception: Additions, alterations, repairs and changes of occupancy to existing buildings complying with Sections N1109 through N1113 as applicable.

N1101.13.1 (R401.2.1) Prescriptive Compliance Option. The Prescriptive Compliance Option requires compliance with Sections N1101.14 through N1104.

N1101.13.2 (R401.2.2) Total Building Performance Option. The Total Building Performance Option requires compliance with Section N1105.

N1101.13.3 (R401.2.3) Energy Rating Index Option. The Energy Rating Index (ERI) Option requires compliance with Section N1106.

N1101.13.4 (R401.2.4) Tropical Climate Region Option. The Tropical Climate Region Option requires compliance with Section N1107.

The prescriptive path sets the maximum *U*-factor for windows.

Note: The Mandatory and Prescriptive labels have been removed from the following section titles:

N1101.14 (R401.3) Certificate ~~(Mandatory)~~.

N1102.1 (R402.1) General ~~(Prescriptive)~~.

N1102.2 (R402.2) Specific insulation requirements ~~(Prescriptive)~~.

N1102.3 (R402.3) Fenestration ~~(Prescriptive)~~.

N1102.4 (R402.4) Air leakage ~~(Mandatory)~~.

N1102.5 (R402.5) Maximum fenestration U-factor and SHGC ~~(Mandatory)~~.

N1103.1 (R403.1) Controls ~~(Mandatory)~~.

N1103.1.2 (R403.1.2) Heat pump supplementary heat ~~(Mandatory)~~.

N1103.3.1 (R403.3.1) ~~Insulation~~ <u>Ducts located outside conditioned space.</u> ~~(Prescriptive)~~.

~~N1103.3.2~~ <u>N1103.3.4</u> (~~R403.3.2~~ <u>R403.3.5</u>) Sealing ~~(Mandatory)~~.

~~N1103.3.3~~ <u>N1103.3.5</u> (~~R403.3.3~~ <u>R403.3.5</u>) Duct testing ~~(Mandatory)~~.

~~N1103.3.4~~ <u>N1103.3.6</u> (~~R403.3.4~~ <u>R403.3.6</u>) Duct leakage ~~(Prescriptive)~~.

~~N1103.3.5~~ <u>N1103.3.7</u> (~~R403.3.5~~ <u>R403.3.7</u>) Building cavities ~~(Mandatory)~~.

N1103.4 (R403.4) Mechanical system piping insulation ~~(Mandatory)~~.

N1103.5.1 (R403.5.1) Heated water circulation and temperature maintenance systems ~~(Mandatory)~~.

~~N1103.5.3~~ <u>N1103.5.2</u> (~~R403.5.3~~ <u>R403.5.2</u>) Hot water pipe insulation ~~(Prescriptive)~~.

N1103.6 (R403.6) Mechanical ventilation ~~(Mandatory)~~.

N1103.7 (R403.7) Equipment sizing and efficiency rating ~~(Mandatory)~~.

N1103.8 (R403.8) Systems serving multiple dwelling units ~~(Mandatory)~~.

N1103.9 (R403.9) Snow melt and ice system controls ~~(Mandatory)~~.

Note: Due to space limitations, not all changes are shown. Please refer to the 2021 IRC for the complete text.

CHANGE SIGNIFICANCE: Beginning in the 2006 code, Section N1101.13 has identified two options for compliance. The options were not formally named within the code but have commonly been referred to as the "prescriptive" and "performance" compliance paths. In addition, a third compliance path was added in 2015, along with a simplified compliance path applicable to certain dwelling units located in the Tropical Zone. The latter compliance option was not listed with the compliance paths.

Section N1101.13 is now titled "Application" for consistency and has been revised to clarify and name the four compliance paths. Naming the compliance options (Prescriptive, Total Building Performance, Energy Rating Index and Tropical Zone) formalizes the way in which the paths are typically identified and adds clarity for training, education and new code users. Using the word "option" reinforces that it is the designer's choice as to which path is followed. The language now clearly leads the code user to the four compliance options.

The terms "Prescriptive" and "Mandatory" in the section titles throughout Sections N1101 through N1104 are eliminated to reduce confusion about the application of the terms. This corresponds to the creation of project requirement tables in Sections N1105 and N1106, and all requirements associated with the respective compliance paths are contained within the associated sections. Additionally, the requirements for Tropical Zone compliance were relocated to the new Section N1107.

The new reference to Section N1101.13.5 directs the user to additional new requirements that apply to all compliance paths and are discussed separately under Sections N1101.13.5 and N1108 in this publication.

N1101.13.5
Additional Energy Efficiency Requirements

CHANGE TYPE: Addition

CHANGE SUMMARY: Additional energy efficiency measures are required regardless of the compliance path used.

2021 CODE: <u>**N1101.13.5 (R401.2.5) Additional energy efficiency.** This section establishes additional requirements applicable to all compliance approaches to achieve additional energy efficiency.</u>

<u>1. For buildings complying with Section N1101.13.1, one of the additional efficiency package options shall be installed according to Section N1108.2.</u>

<u>2. For buildings complying with Section N1101.13.2 the building shall meet one of the following:</u>
   <u>2.1. One of the additional efficiency package options in Section N1108.2 shall be installed without including such measures in the proposed design under Section N1105.</u>
   <u>2.2. The proposed design of the building under Section N1105.3 shall have an annual energy cost that is less than or equal to 95 percent of the annual energy cost of the standard reference design.</u>

Enhanced envelope performance is one option for satisfying the additional efficiency requirements for the prescriptive or performance path.

3. For buildings complying with Section N1101.13.3, the Energy Rating Index value shall be at least 5 percent less than the Energy Rating Index target specified in Table N1106.5.

The option selected for compliance shall be identified on the certificate required by Section N1101.14.

CHANGE SIGNIFICANCE: Arguably the most significant change to Chapter 11 of the IRC and the residential provisions of the 2021 IECC, new Sections N1101.13.5 and N1108 set additional requirements to conserve energy. Whichever compliance path is chosen, one of the additional options must also be used.

Section N1101.13.5 establishes the application of the additional requirements:

- Prescriptive compliance projects must include one package from Section N1108.
- Total building performance compliance projects must either include one package from Section N1108, without including the additional measures in the proposed design, or the proposed design must have an annual energy cost less than or equal to 95 percent of the standard reference design.
- Projects complying via the Energy Rating Index must have a score at least 5 percent less than the maximum value shown in Table N1106.5.

The additional energy efficiency package options are covered under Section N1108 in this publication.

N1101.14
Permanent Energy Certificate

CHANGE TYPE: Modification

CHANGE SUMMARY: Additional information related to the building thermal envelope, solar energy, Energy Rating Index and the code edition is required on the energy certificate.

2021 CODE: N1101.14 (R401.3) Certificate (Mandatory). A permanent certificate shall be completed by the builder or other approved party and posted on a wall in the space where the furnace is located, a utility room or an approved location inside the building. Where located on an electrical panel, the certificate shall not cover or obstruct the visibility of the circuit directory label, service disconnect label or other required labels. The certificate shall indicate <u>the following:</u>

1. The predominant R-values of insulation installed in or on ceilings, roofs, walls, foundation components such as slabs, basement walls, crawl space walls and floors, and ducts outside conditioned spaces.

2. <u>The</u> U-factors of fenestration and the solar heat gain coefficient (SHGC) of fenestration<u>.</u>, and Where there is more than one value for each <u>any</u> component <u>of the building envelope</u>, the certificate shall indicate <u>both</u> the value covering the largest area <u>and the area weighted average value if available</u>.

Energy Efficiency Certificate

Insulation rating	R-Value			R-Value
Ceiling/Roof	_____		Floor/Foundation	_____
Wall	_____		Ductwork	_____

Glass & door rating	U-Factor	SHGC	U-Factor	SHGC
Window	_____	_____	_____	_____
Door	_____	_____	_____	_____

Heating & cooling equipment		Efficiency
Heating system:	_____	_____
Cooling system:	_____	_____
Water heater:	_____	_____

Building air leakage and duct test results			
Building air leakage	_____	Name of tester	_____
Duct test	_____	Name of tester	_____

Photovoltaic (PV) panel system			
Array capacity	_____	Panel tilt	_____
Inverter efficiency	_____	Orientastion	_____

Energy Rating Index (ERI)			
ERI w/o on-site generation	_____	ERI with on-site generation	_____

Additional energy efficiency option used: _____

Name: _____ Date: _____

Sample energy certificate.

3. The results from any required duct system and building envelope air leakage testing performed on the building.

4. The ~~certificate shall indicate the~~ types, sizes and efficiencies of heating, cooling and service water heating equipment. Where a gas-fired unvented room heater, electric furnace, or baseboard electric heater is installed in the residence, the certificate shall indicate "gas-fired unvented room heater," "electric furnace" or "baseboard electric heater," as appropriate. An efficiency is not required to ~~shall not~~ be indicated for gas-fired unvented room heaters, electric furnaces and electric baseboard heaters.

5. Where onsite photovoltaic panel systems have been installed, the array capacity size, inverter efficiency, panel tilt and orientation shall be noted on the certificate.

6. For buildings where an Energy Rating Index score is determined in accordance with Section N1106, the Energy Rating Index score, both with and without any on-site generation, shall be listed on the certificate.

7. The code edition under which the structure was permitted and the compliance path used.

CHANGE SIGNIFICANCE: The IRC energy provisions require the builder or other approved party to complete a permanent energy efficiency certificate and post it near the furnace or other approved location. The certificate lists the primary energy conserving features of the building, the results of air leakage testing and the efficiency ratings of installed heating, cooling and water heating equipment. Providing this information in a permanent location benefits current and future homeowners and certifies that the building complies with the energy efficiency requirements in effect at the time of construction. It is important to record this information during construction as the data can be difficult or impossible to recreate later when needed for maintenance or real estate transaction purposes.

The requirement for posting a permanent certificate first appeared in the energy efficiency provisions of the 2006 IRC and have been modified in each edition of the code since then to reflect trends in energy conservation efforts. For example, results of the blower door test for building air leakage and the duct system air test were added to the certificate in the 2009 IRC. In the 2015 edition, the posting location changed to assure the information was visible to future homeowners. The 2021 IRC requires additional details on the building thermal envelope efficiency (where available) and HVAC equipment size. Reflecting the national trend toward more solar energy systems, specifications for any onsite photovoltaic (PV) panel systems must also appear on the certificate. If the optional compliance path of an energy rating index is used, the resulting score must be listed on the certificate. Finally, the code edition used for compliance with the energy efficiency provisions must be identified on the permanent certificate.

N1102.1
Building Thermal Envelope

CHANGE TYPE: Modification

CHANGE SUMMARY: The assembly *U*-Factor is established as the primary insulation metric. The *R*-Value approach is now an alternative method.

2021 CODE: N1102.1 (R402.1) General ~~(Prescriptive)~~. The building thermal envelope shall comply with the requirements of Sections N1102.1.1 through N1102.1.5.

Exceptions: [No changes]

N1102.1.1 (R402.1.1) Vapor retarder. Wall assemblies in the building thermal envelope shall comply with the vapor retarder requirements of Section R702.7.

N1102.1.2 (R402.1.2) Insulation and fenestration criteria. The building thermal envelope shall meet the requirements of Table N1102.1.2 based on the climate zone specified in Section N1101.7. <u>Assemblies shall have a *U*-factor equal to or less than that specified in Table N1102.1.2. Fenestration shall have a *U*-factor and glazed fenestration SHGC equal to or less than specified in Table N1102.1.2.</u>

<u>N1102.1.3 (R402.1.3) *R*-value alternative.</u> <u>Assemblies with *R*-value of insulation materials equal to or greater than that specified in Table N1102.1.3 shall be an alternative to the *U*-factor in Table N1102.1.2.</u>

~~N1102.1.3 (R402.1.3)~~ <u>N1102.1.4 (R402.1.4)</u> *R*-value computation. ~~Insulation material used in layers, such as framing cavity insulation or continuous insulation,~~ <u>Cavity insulation alone shall be used to determine compliance with the cavity insulation *R*-value requirements in</u>

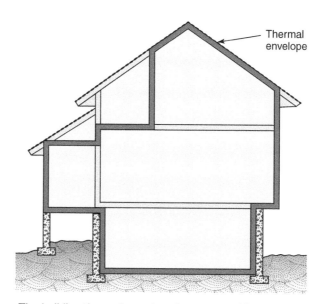

The building thermal envelope is an assembly of elements that provide a boundary between conditioned space and unconditioned space.

<u>Table N1102.1.3. Where cavity insulation is installed in multiple layers, the
R -values of the cavity insulation layers</u> shall be summed to ~~compute
the corresponding component R-value~~ <u>determine compliance with the
cavity insulation R -value requirements.</u> The manufacturer's settled
R-value shall be used for blown-in insulation. <u>Continuous insulation (ci)
alone shall be used to determine compliance with the continuous insulation R -value requirements in Table N1102.1.3. Where continuous
insulation is installed in multiple layers, the R-values of the continuous insulation layers shall be summed to determine compliance with the
continuous insulation R -value requirements. Cavity insulation R -values
shall not be used to determine compliance with the continuous insulation
R-value requirements in Table N1102.1.3.</u> Computed R-values shall
not include an R-value for other building materials or air films. Where
insulated siding is used for the purpose of complying with the continuous
insulation requirements of Table ~~N1102.1.2~~ <u>N1102.1.3,</u> the manufacturer's
labeled R-value for insulated siding shall be reduced by R-0.6.

CHANGE SIGNIFICANCE: Section N1102 provides prescriptive building thermal envelope provisions, including limitations on air infiltration
(leakage). The building envelope is important to building energy efficiency. When it is cold outside, heat loss and air leakages through the
building envelope add to the heating load. On hot days, solar gains
through windows, heat gain through opaque assemblies and infiltration
of hot or humid air contribute to the cooling load. The building envelope requirements of Section N1102 are intended to reduce heat gains and
losses through the building envelope.

The building envelope provisions have seen few significant changes
since the 2012 edition of the code. Significant changes to the building
envelope in the 2021 IRC include:

- Increased insulation requirements
- Reduced fenestration U-factors and solar heat gain coefficients (SHGC)
- Increased resistance to air leakage, including sealing of electrical boxes

Since the 2006 editions of the IECC and IRC, R-values have been the
primary method for measuring envelope efficiencies, with assembly
U-factors an alternative compliance metric. The change to Section
N1102.1 flips the focus of envelope efficiency to U-factors and organizes
the envelope thermal requirements such that the basis of performance
(and any other means of compliance) is founded on U-factors which completely define an assembly's performance. The R-value approach is
retained, providing predetermined solutions based on the U-factor
requirements. This reorientation ensures that the R-values used are a de-
rivative of the intended performance levels that are nonmaterial specific
and represented by the assembly U-factors. The SHGC values (and footnotes) are consistent with those in the revised R-value table.

Tables N1102.1.2 and N1102.1.3
Insulation and Fenestration Requirements

CHANGE TYPE: Modification

CHANGE SUMMARY: Increased energy efficiency is achieved with lowered assembly U-factor requirements in Table N1102.1.2 and increased R-value requirements in Table N1102.1.3.

2021 CODE:

TABLE ~~N1102.1.4 (R402.1.4)~~ <u>N1102.1.2 (R402.1.2)</u> ~~Equivalent~~ <u>Maximum Assembly U-Factors[a] and Fenestration Requirements</u>

Climate Zone	Fenestration U-Factor[f]	Skylight U-Factor	Glazed Fenestration SHGC[d,e]	Ceiling U-Factor	Frame Wall U-Factor	Mass Wall U-Factor[b]	Floor U-Factor	Basement Wall U-Factor	Crawl Space Wall U-Factor
<u>0</u>	<u>0.50</u>	<u>0.75</u>	<u>0.25</u>	<u>0.035</u>	<u>0.084</u>	<u>0.197</u>	<u>0.064</u>	<u>0.360</u>	<u>0.477</u>
1	0.50	0.75	<u>0.25</u>	0.035	0.084	0.197	0.064	0.360	0.477
2	0.40	0.65	<u>0.25</u>	~~0.030~~ <u>0.026</u>	0.084	0.165	0.064	0.360	0.477
3	~~0.32~~ <u>.30</u>	0.55	<u>0.25</u>	~~0.030~~ <u>0.026</u>	0.060	0.098	0.047	0.091[c]	0.136
4 except Marine	~~0.32~~ <u>.30</u>	0.55	0.40	~~0.026~~ <u>0.024</u>	~~0.060~~ <u>0.045</u>	0.098	0.047	0.059	0.065
5 and Marine 4	0.30	0.55	<u>NR</u>	~~0.026~~ <u>0.024</u>	~~0.060~~ <u>0.045</u>	0.082	0.033	0.050	0.055
6	0.30	0.55	<u>NR</u>	~~0.026~~ <u>0.024</u>	0.045	0.060	0.033	0.050	0.055
7 and 8	0.30	0.55	<u>NR</u>	~~0.026~~ <u>0.024</u>	0.045	0.057	0.028	0.050	0.055

a. Nonfenestration U-factors shall be obtained from measurement, calculation or an approved source.
b. Mass walls shall be in accordance with Section N1102.2.5. Where more than half the insulation is on the interior, the mass wall U-factors shall not exceed 0.17 in Climate Zone 0 and 1, 0.14 in Climate Zone 2, 0.12 in Climate Zone 3, 0.087 in Climate Zone 4 except Marine, 0.065 in Climate Zone 5 and Marine 4, and 0.057 in Climate Zones 6 through 8.
c. In Warm-Humid locations as defined by Figure N1101.7 and Table N1101.7, the basement wall U-factor shall not exceed 0.360.
d. <u>The SHGC column applies to all glazed fenestration.</u>

> **Exception:** <u>In Climate Zones 0 through 3, skylights shall be permitted to be excluded from glazed fenestration SHGC requirements provided that the SHGC for such skylights does not exceed 0.30</u>

e. <u>There are no SHGC requirements in the Marine Zone.</u>
f. <u>A maximum U-factor of 0.32 shall apply in Marine Climate Zone 4 and Climate Zones 5 through 8 to vertical fenestration products installed in buildings located in accordance with one of the following:</u>
   1. <u>Above 4000 feet in elevation above sea level.</u>
   2. <u>In windborne debris regions where protection of openings is required by Section R301.2.1.2.</u>

~~TABLE N1102.1.2 (R402.1.2)~~ <u>TABLE N1102.1.3 (R402.1.3)</u> Insulation <u>Minimum R-Values</u> and Fenestration Requirements by Component[a]

Climate Zone	Fenestration U-Factor[b,i]	Skylight[b] U-Factor	Glazed Fenestration SHGC[b,e]	Ceiling R-Value	Wood Frame Wall R-value[hg]	Mass Wall R-value[ih]	Floor R-Value	Basement[c,d] Wall R-Value	Slab[e] R-value & Depth	Crawl Space[c,d] Wall R-value
0	NR	0.75	0.25	30	13 or 0 + 10ci	3/4	13	0	0	0
1	NR	0.75	0.25	30	13 or 0 + 10ci	3/4	13	0	0	0
2	0.40	0.65	0.25	~~38~~ <u>49</u>	13 or 0 + 10ci	4/6	13	0	0	0
3	~~0.32~~ <u>0.30</u>	0.55	0.25	~~38~~ <u>49</u>	20 or 13 + 5ci or 0 + 15ci	8/13	19	5<u>ci</u> ~~/~~<u>or</u> 13[g]	~~0~~ <u>10ci,</u> 2 ft	5<u>ci</u> ~~/~~<u>or</u> 13
4 except Marine	~~0.32~~ <u>0.30</u>	0.55	0.40	~~49~~ <u>60</u>	20 or 13 + 5ci or 0 + 15ci	8/13	19	10<u>ci</u> ~~/~~<u>or</u> 13	10<u>ci</u>, ~~2~~ <u>4</u> ft	10 <u>ci</u> ~~/~~<u>or</u> 13
5 and Marine 4	0.30	0.55	NR	~~49~~ <u>60</u>	20 or 13 + 5ci or 0 + 15ci	13/17	30[g]	15<u>ci</u> ~~/~~<u>or</u> 19 or 13+5ci	10<u>ci</u>, ~~2~~ <u>4</u> ft	15<u>ci</u> ~~/~~ or 19 or 13+5ci
6	0.30	0.55	NR	~~49~~ <u>60</u>	30 or 20 + 5ci or 13 + 10ci or 0+20ci	15/20	30[g]	15<u>ci</u> ~~/~~<u>or</u> 19 or 13+5ci	10<u>ci</u>, 4 ft	15<u>ci</u> ~~/~~ or 19 or 13+5ci
7 and 8	0.30	0.55	NR	~~49~~ <u>60</u>	30 or 20 + 5ci or 13 + 10ci or 0+20ci	19/21	30[g]	15<u>ci</u> ~~/~~<u>or</u> 19 or 13+5ci	10<u>ci</u>, 4 ft	15<u>ci</u> ~~/~~ or 19 or 13+5ci

For SI:1 foot = 304.8 mm.

a. R-values are minimums. U-factors and SHGC are maximums. Where insulation is installed in a cavity that is less than the label or design thickness of the insulation, the installed R-value of the insulation shall be not less than the R-value specified in the table.
b. The fenestration U-factor column excludes skylights. The SHGC column applies to all glazed fenestration.

   Exception: In Climate Zones ~~1~~ <u>0</u> through 3, skylights shall be permitted to be excluded from glazed fenestration SHGC requirements provided that the SHGC for such skylights does not exceed 0.30.

c. "5<u>ci</u> or 13" means R-5 continuous insulation (ci) on the interior or exterior surface of the wall or R-13 cavity insulation on the interior side of the wall. "10<u>ci</u> ~~/~~<u>or</u> 13" means R-10 continuous insulation <u>(ci)</u> on the interior or exterior <u>surface</u> of the ~~home~~ <u>wall</u> or R-13 cavity insulation ~~at~~ <u>on</u> the interior <u>side</u> of the ~~basement~~ wall. "15<u>ci</u> ~~/~~<u>or</u> 19" means R-15 continuous insulation on the interior or exterior <u>surface</u> of the ~~home~~ <u>wall</u> or R-19 cavity insulation on the interior <u>side</u> of the ~~basement~~ wall. ~~Alternatively, compliance with "15/19" shall be~~ "13+5ci" means R-13 cavity insulation on the interior <u>side</u> of the ~~basement~~ wall ~~plus~~ <u>in addition to</u> R-5 continuous insulation on the interior or exterior <u>surface</u> of the ~~home~~ <u>wall</u>.
d. R-5 insulation shall be provided under the full slab area of a heated slab in addition to the required slab edge R-value for slabs, as indicated in the table. The slab edge insulation for heated slabs shall not be required to extend below the slab.
e. There are no SHGC requirements in the Marine Zone.
f. Basement wall insulation shall not be required in warm-humid locations as defined by Figure ~~N1101.10~~ <u>N1101.7</u> and Table ~~N1101.10~~ <u>N1101.7</u>.
~~g. Alternatively, insulation sufficient to fill the framing cavity providing not less than an R-value of R-19.~~
~~h.~~ <u>g.</u> The first value is cavity insulation, the second value is continuous insulation. Therefore, as an example, "13+5<u>ci</u>" means R-13 cavity insulation plus R-5 continuous insulation.
~~i.~~ <u>h.</u> Mass walls shall be in accordance with Section N1102.2.5. The second R-value applies where more than half of the insulation is on the interior of the mass wall.
i. <u>A maximum U-Factor of 0.32 shall apply in Climate Zones 3 through 8 to vertical fenestration products installed in buildings located either:
   1. Above 4000 feet (1220 m) in elevation, or
   2. In windborne debris regions where protection of openings is required by Section R301.2.1.2.</u>

CHANGE SIGNIFICANCE: The order of the two tables controlling the prescriptive approach for energy efficiency of the building thermal envelope has been reversed to recognize that the assembly U-factors are the primary method for determining compliance (as discussed in the previous change to Section N1102.1). Some assembly U-factors in Table N1102.1.2 have been lowered and corresponding R-value requirements in Table N1102.1.4 have increased.

In particular, efficiencies are increased in ceiling assemblies, wood frame walls and fenestration, depending on climate zone. Ceiling requirements have increased in all climate zones except Climate Zone 1. For example, the minimum R-value for ceilings in Climate Zones 4 through 8 have increased from R49 to R60 (assembly U-Factors decreased from 0.026 to 0.024). Wood frame wall efficiencies are more stringent in Climate Zones 4 and 5. Fenestration U-factors in Climate Zones 3 and 4 are decreased to improve efficiency. An SHGC value has been introduced in climate zone 5 where it was previously not required.

Slab-on-grade insulation has been increased in efficiency for Climate Zones 3 through 5. This is considered an overdue update because the slab R-value requirements have not improved in any climate zone since at least 2006.

Other changes relate to footnotes, additional options for using continuous insulation (ci), and clarification to the tables. Basement and crawlspace walls are amended for clarity and move the alternative compliance option (13+5ci) from a footnote to the table. The clarification is made by adding a "ci" designator for the continuous insulation components in the table. This is also carried through in the footnote "c". Footnote "c" is further clarified by consistently referring to "wall" rather than in some cases "basement" and in other cases "home" with no mention of "crawl space." The footnote applies to basement and crawl space walls as noted in the heading of the table columns. Also, the word "surface" is added to clarify that where continuous insulation is used it should be applied to the surface of the foundation wall (with any additional cavity insulation applied to the interior side of the continuous insulation).

The elimination of footnote g related to floor insulation in Climate Zones 4 through 8 improves the efficiency of dwellings in the coldest climate zones by removing an exception that allowed lower floor insulation R-values with no corresponding improvements elsewhere in the building.

Energy efficiency for ceilings has increased in most climate zones.

A footnote is also established, providing an exception to prescriptive *U*-factors for fenestration installed at high altitudes (above 4000 feet in elevation) and in regions that require fenestration to be resistant to wind-borne debris in Climate Zones 3 through 8.

N1102.2
Ceiling Insulation

CHANGE TYPE: Modification

CHANGE SUMMARY: The options for a reduction in *R*-values for both ceilings with attics and those without have been adjusted to recognize the increase in the prescriptive ceiling *R*-values in Table N1102.1.3.

2021 CODE: N1102.2 (R402.2) Specific insulation requirements (Prescriptive). In addition to the requirements of Section N1102.1, insulation shall meet the specific requirements of Sections N1102.2.1 through N1102.2.12.

N1102.2.1 (R402.2.1) Ceilings with attics spaces. Where Section R1102.1.2 R1102.1.3 requires R-38 R-49 insulation in the ceiling or attic, installing R-30 R-38 insulation over 100 percent of the ceiling or attic area requiring insulation shall satisfy the requirement for R-38 R-49 insulation wherever the full height of uncompressed R-30 R-38 insulation extends over the wall top plate at the eaves. Where Section N1102.1.2 N1102.1.3 requires R-49 R-60 insulation in the ceiling or attic, installing R-38 R-49 over 100 percent of the ceiling or attic area requiring insulation shall satisfy the requirement for R-49 R-60 insulation wherever the full height of uncompressed R-38 R-49 insulation extends over the top plate at the eaves. This reduction shall not apply to the insulation and fenestration criteria in U-factor alternative approach in Section N1102.1.4 N1102.1.3 and the Total UA alternative in Section N1102.1.5.

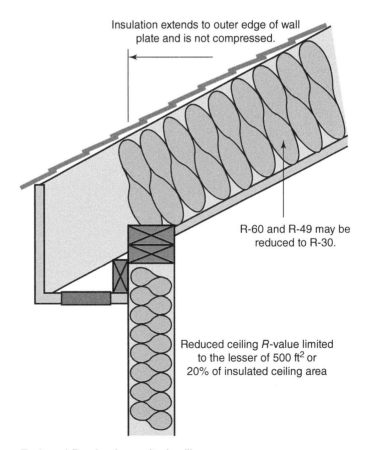

Reduced *R*-value for vaulted ceiling.

N1102.2.2 (R402.2.2) Ceilings without attic~~s~~ spaces. Where Section ~~N1102.1.2~~ <u>N1102.1.3</u> requires insulation *R*-values greater than R-30 in the ~~ceiling~~ <u>interstitial space above a ceiling and below the structural roof deck</u> and the design of the roof/ceiling assembly does not allow sufficient space for the required insulation, the minimum required insulation *R*-value for such roof/ceiling assemblies shall be R-30. Insulation shall extend over the top of the wall plate to the outer edge of such plate and shall not be compressed. This reduction of insulation from the requirements of Section ~~N1102.1.2~~ <u>N1102.1.3</u> shall be limited to 500 square feet (46 m²) or 20 percent of the total insulated ceiling area, whichever is less. This reduction shall not apply to the ~~*U*-factor alternative approach in Section N1102.1.4 and the~~ Total UA alternative in Section N1102.1.5.

N1102.2.3 (R402.2.3) Eave baffle. For air-permeable insulations in vented attics, a baffle shall be installed adjacent to soffit and eave vents. Baffles shall maintain ~~an~~ <u>a net free area</u> opening equal or greater than the size of the vent. The baffle shall extend over the top of the attic insulation. The baffle shall be permitted to be any solid material. <u>The baffle shall be installed to the outer edge of the exterior wall top plate so as to provide maximum space for attic insulation coverage over the top plate. Where soffit venting is not continuous, baffles shall be installed continuously to prevent ventilation air in the eave soffit from bypassing the baffle.</u>

CHANGE SIGNIFICANCE: In the revised prescriptive tables, the minimum *R*-value for ceilings in Climate Zones 4 through 8 have increased from R-49 to R-60 (assembly *U*-factors decreased from 0.026 to 0.024). In Climate Zones 2 and 3, the ceiling *R*-value increased from R-38 to R-49 (assembly *U*-factors decreased from 0.030 to 0.026). The update to Sections N1102.2.1 and N1102.2.2 recognizes those changes and factors in the increase to the prescriptive values in the referenced tables.

The IRC has long offered two exceptions to the prescriptive ceiling insulation requirements. The first in Section N1102.2.1, Ceilings with Attic Spaces, recognizes the effectiveness of energy (raised heel) trusses. The increased height at the exterior wall plate allows for the full thickness of ceiling insulation throughout and the code accordingly allows a reduction in the required *R*-value. The second exception in Section N1102.2.2, Ceilings without Attic Spaces, recognizes the difficulty of achieving the full required thickness of insulation in vaulted or cathedral ceilings framed with solid-sawn dimension lumber, or perhaps I-joists, rather than trusses. In this case, the code allows a reduction in *R*-value but limits the total area for the reduction.

The reductions have been revised to reflect the changes to the prescriptive tables and are based on the prescriptive value for the particular climate zone. For houses with attics, a prescriptive *R*-value of R-60 may be reduced to R-49 for the entire area provided energy (raised heel) trusses are used to allow the full height of insulation above the wall top plate. Compression of the insulation at any point reduces the *R*-value and is less effective at conserving energy.

For ceilings without attics, the reduction in *R*-value is limited to 500 square feet or 20 percent of the total insulated ceiling area, whichever is less. Again, the full depth of the insulation must continue to the outside of the wall top plate. Where the prescriptive ceiling value is greater than R-30, the vaulted ceiling value can be reduced to R-30.

In a clarification to Section N1102.2.3, the requirements have been strengthened to ensure optimal performance of the eave baffles in providing proper circulation of ventilation air.

N1102.2.4 Access Hatches and Doors

CHANGE TYPE: Modification

CHANGE SUMMARY: Prescriptive provisions for insulation levels and mandatory provisions for installation (weather-stripping and baffles) have been placed in separate sections. Provisions addressing pull-down stairs have been added.

2021 CODE: N1102.2.4 (R402.2.4) Access hatches and doors. Access hatches and doors from conditioned to unconditioned spaces such as attics and crawlspaces shall be insulated to the same level required for the wall or ceiling R-value in Table N1102.1.3 in which they are installed.

Exceptions:

1. Vertical doors providing access from conditioned spaces to unconditioned spaces that comply with the fenestration requirements of Table N1102.1.3 based on the applicable climate zone specified in Section N1101.7.

2. Horizontal pull-down, stair-type access hatches in ceiling assemblies that provide access from conditioned to unconditioned spaces in Climate Zones 0 through 4 shall not be required to comply with the insulation level of the surrounding surfaces provided that the hatch meets all of the following:

    2.1. The average U-factor of the hatch shall be less than or equal to U-0.10 or have an average insulation R-value of R-10 or greater.

    2.2. Not less then 75 percent of the panel area shall have an insulation R-value of R-13 or greater.

    2.3. The net area of the framed opening shall be less than or equal to 13.5 square feet (1.25 m^2).

    2.4. The perimeter of the hatch edge shall be weatherstripped.

The reduction shall not apply to the total UA alternative in Section N1102.1.5.

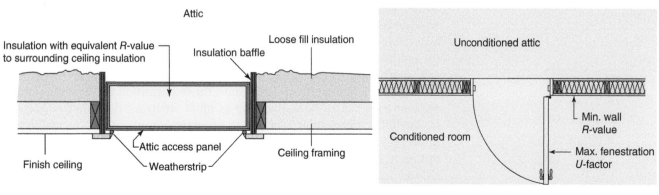

Horizontal attic access

Vertical attic access

Approved vertical and horizontal access to unconditioned attics.

N1102.2.4 (R402.2.4) **N1102.2.4.1 (R402.2.4.1) Access** ~~hatches~~ **hatch and** ~~doors~~ **door insulation installation and retention.** ~~Access~~ Vertical or horizontal access hatches and doors from conditioned spaces to unconditioned spaces such as attics and crawl spaces shall be weatherstripped ~~and insulated to a level equivalent to the insulation on the surrounding surfaces~~. Access that prevents damaging or compressing the insulation shall be provided to all equipment. Where loose-fill insulation is installed, a wood-framed or equivalent baffle, ~~or~~ retainer or dam shall be installed to prevent ~~the~~ loose-fill insulation from spilling into ~~the~~ living spaces ~~when the attic access is opened~~, from higher to lower sections of the attic, and from attics covering conditioned spaces to unconditioned spaces.. The baffle or retainer shall provide a permanent means of maintaining the installed R-value of the loose-fill insulation.

> **Exception:** ~~Vertical doors providing access from conditioned spaces to unconditioned spaces that comply with the fenestration requirements of Table N1102.1.2 based on the applicable climate zone specified in Section N1101.7.~~

CHANGE SIGNIFICANCE: Section N1102.2.4 regulates the installation of access panels and doors entering unconditioned spaces such as attics. The provisions are mandatory for the prescriptive, total building performance and ERI compliance paths. To conserve energy, this section sets minimum levels of insulation for the panel as part of the building thermal envelope based on climate zone. Installation requirements include weather-stripping to reduce air leakage and baffles to prevent displacement of insulation.

The approach for horizontal attic hatches has been to insulate the hatch to meet the required R-value of the surrounding area of the attic. In the 2015 edition of the code, it was recognized that insulating vertical doors to the same R-value as the wall was not practical and not consistent with how fenestration, including doors, in the building thermal envelope were treated in other situations. Subsequently, vertical access panels or doors were treated as fenestration and only needed to meet the associated prescriptive U-factor requirements. Similarly, in the 2021 edition, an exception is added to allow pull-down stair-type access hatches in ceiling assemblies for transitioning from conditioned to unconditioned space. This exception is limited to Climate Zones 0 through 4 and allows a reduced R-value (or increased U-factor). The area of the pull-down stair is limited to 13.5 square feet. The reason behind the change was that horizontal access panels were subjected to a higher level of performance than what is required for skylights in an insulated ceiling assembly and that affordable, premanufactured pull-down stair access systems are not readily available to meet the R-30 to R-49 requirement. The U-factor specified at U-0.10 is less stringent than the U-factors specified for the insulated ceilings but is far more stringent than those permitted for skylights in all climate zones. Also, the size limit is more stringent than that permitted for skylights. Consensus was that the practical implications outweigh the minimal loss of insulation R-value in this application and that having this option in the code might discourage construction of "field-crafted detachable apparatuses" that were often discarded.

N1102.2.7
R-Value Reduction for Walls with Partial Structural Sheathing

CHANGE TYPE: Deletion

CHANGE SUMMARY: The provision for reducing the *R*-value of the required continuous wall insulation at areas of structural wall sheathing has been deleted.

2021 CODE: ~~**N1102.2.7 (R402.2.7) Walls with partial structural sheathing.** Where Section N1102.1.2 requires continuous insulation on exterior walls and structural sheathing covers 40 percent or less of the gross area of all exterior walls, the required continuous insulation *R*-value shall be permitted to be reduced by an amount necessary, but not more than R-3, to result in a consistent total sheathing thickness on areas of the walls covered by structural sheathing. This reduction shall not apply to the *U*-factor alternative in Section N1102.1.4 and the Total UA alternative in Section N1102.1.5.~~

CHANGE SIGNIFICANCE: Previously, the code had allowed a compromise to balance structural requirements with energy efficiency requirements as it relates to the use of wood structural panels for wall bracing. In colder climates, it is often necessary to use continuous exterior wall insulation in conjunction with cavity insulation. The code allowed a reduced *R*-value for the continuous insulation for an area up to 40 percent

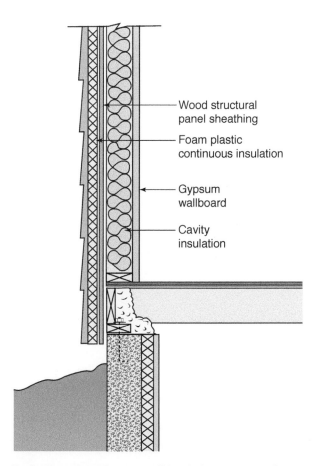

Reduction of continuous wall insulation at areas of structural wall bracing is no longer permitted.

of the total exterior wall area. This was a way to achieve uniform total thickness of wall sheathing throughout – the full thickness of insulation over 60 percent of the walls would be equal to the reduced thickness of insulation plus the thickness of the wall bracing over 40 percent of the area. In the interest of improving energy efficiency in homes, this option has been deleted as being unnecessary. Unlike other trade-offs in the code, this prescriptive exception did not require any additional energy savings elsewhere to offset the loss in efficiency. Consensus for this change was based on the concept that the prescriptive thermal envelope requirements were reasonable and that there were a broad range of other energy-neutral trade-offs that could be implemented in the design of the building.

N1102.2.7
Floor Insulation

CHANGE TYPE: Clarification

CHANGE SUMMARY: The floor insulation provisions have been placed into three separate methods of compliance for clarification.

2021 CODE: ~~N1102.2.8 (R402.2.8)~~ <u>N1102.2.7 (R402.2.7)</u> **Floors.** Floor ~~framing~~ cavity insulation shall <u>comply with one of the following:</u>

1. <u>Insulation shall</u> be installed to maintain permanent contact with the underside of the subfloor decking <u>in accordance with manufacturer instructions to maintain required R-value or readily fill the available cavity space.</u>

2. <u>Floor framing cavity insulation shall be permitted to be in contact with the top side of sheathing separating the cavity and the unconditioned space below. Insulation shall extend from the bottom to the top of all perimeter floor framing members and the framing members shall be air sealed.</u>

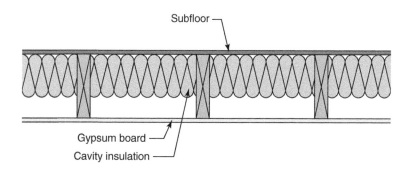

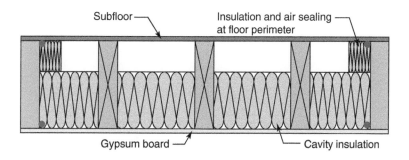

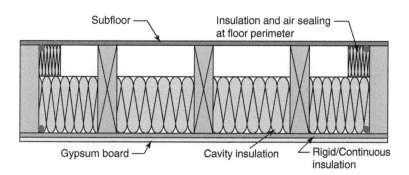

Three options for floor cavity insulation.

3. A combination of cavity and continuous insulation shall be installed so that the cavity insulation is **Exception:** ~~As an alternative, the floor framing-cavity insulation shall be~~ in contact with the top side of ~~sheathing or~~ the continuous insulation that is installed on the ~~bottom side~~ underside of the floor framing ~~where combined with insulation that meets or exceeds the minimum wood frame wall R-value in Table N1102.1.2 and that extends~~ separating the cavity and the unconditioned space below. The combined *R*-value of the cavity and continuous insulation shall equal the required *R*-value for floors. Insulation shall extend from the bottom to the top of all perimeter floor framing members and the framing members shall be air sealed.

CHANGE SIGNIFICANCE: Prior to the 2015 code, cavity insulation in floor assemblies above unconditioned spaces had to be in contact with the underside of the floor sheathing. In the 2015 edition, another option permitted an air space between the floor sheathing and the top of the cavity insulation. In this case, the cavity insulation is in direct contact with the topside of the sheathing or continuous insulation installed on the underside of the floor framing and is combined with perimeter insulation that meets or exceeds the *R*-value requirements for walls. This second option was seen as having fewer cold spots and did not increase heat loss. It also facilitates ductwork and piping to be enclosed within the thermal envelope.

This second option was written as an exception and the language had been unclear to some code users. For example, the exception mentioned insulation being in contact with the ceiling membrane below or in contact with continuous insulation, raising the question as to whether continuous insulation was required to use this exception. The revised language does not intend to change the requirement, only to clarify and simplify the language. The text is now broken into a list of three separate options to comply with the code.

N1102.2.8 Unconditioned Basement

CHANGE TYPE: Modification

CHANGE SUMMARY: A number of specific criteria must be met for a space to qualify as an unconditioned basement.

2021 CODE: N1102.2.8 (R402.2.8) Basement Walls. Basement walls shall be insulated in accordance with Table N1102.1.3.

Exception: Basement walls associated with unconditioned basements where all of the following requirements are met:
1. The floor overhead, including the underside stairway stringer leading to the basement, is insulated in accordance with Section N1102.1.3 and applicable provisions of Sections N1102.2 and N1102.2.7.
2. There are no uninsulated duct, domestic hot water, or hydronic heating surfaces exposed to the basement.
3. There are no HVAC supply or return diffusers serving the basement.
4. The walls surrounding the stairway and adjacent to conditioned space are insulated in accordance with Section N1102.1.3 and applicable provisions of Section N1102.2.

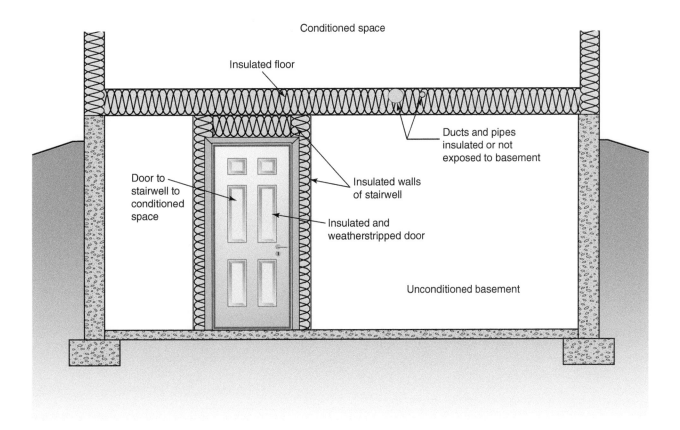

Unconditioned basement requirements.

5. The door(s) leading to the basement from conditioned spaces are insulated in accordance with Section N1102.1.3 and applicable provisions of Section N1102.2, and weather-stripped in accordance with Section N1102.4.

6. The building thermal envelope separating the basement from adjacent conditioned spaces complies with Section N1102.4.

~~N1102.2.9 (R402.2.9)~~ N1102.2.8.1 (R402.2.8.1) Basement walls insulation installation. ~~Walls associated with conditioned basements~~ Where basement walls are insulated, the insulation shall be ~~insulated~~ installed from the top of the basement wall down to 10 feet (3048 mm) below grade or to the basement floor, whichever is less. ~~Walls associated with unconditioned basements shall comply with this requirement except where the floor overhead is insulated in accordance with Sections N1102.1.2 and N1102.2.8.~~

CHANGE SIGNIFICANCE: The changes to this section comprehensively clarify what qualifies as an unconditioned basement. Significantly more than insulating the floor over the basement is required to consider the basement unconditioned and not just an extension of conditioned space. Under previous energy codes, heat transfer through ducts, stairways, doors and other elements of the basement were not taken into consideration.

The new language identifies the characteristics that must be present in order for the basement to qualify as an unconditioned space. For compliance with the prescriptive option, basements without these characteristics require wall insulation in accordance with Table N1102.1.3.

N1102.4 and Table N1102.4.1.1
Building Air Leakage and Testing

CHANGE TYPE: Modification

CHANGE SUMMARY: The building air leakage and testing requirements are revised in Section N1102.4.

2021 CODE: <u>**N1101.6 (R202) Dwelling unit enclosure area.** The sum of the area of ceiling, floors, and walls separating a dwelling unit's conditioned space from the exterior or from adjacent conditioned or unconditioned spaces. Wall height shall be measured from the finished floor of the dwelling unit to the underside of the floor above.</u>

N1102.4 (R402.4) Air leakage ~~(Mandatory)~~. The building thermal envelope shall be constructed to limit air leakage in accordance with the requirements of Sections N1102.4.1 through N1102.4.5.

N1102.4.1.2 (R402.4.1.2) Testing. The building or dwelling unit shall be tested ~~and verified as having an air leakage rate of not exceeding five air changes per hour in Climate Zones 1 and 2, and three air changes per hour in Climate Zones 3 through 8~~ <u>for air leakage. The maximum air leakage rate for any building or dwelling unit under any compliance path shall</u>

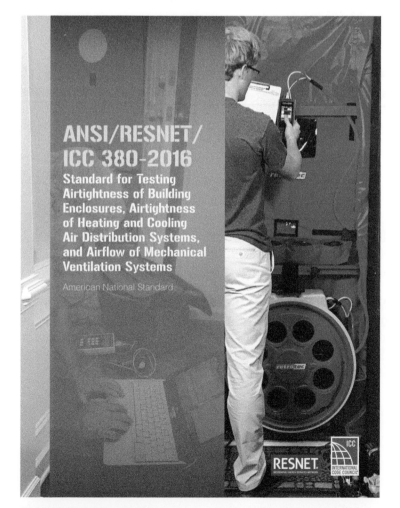

RESNET/ICC 380 establishes test procedures for determining air tightness of buildings.

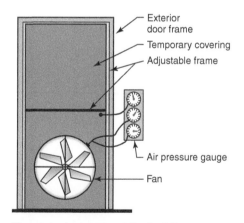

The air leakage rate of the building thermal envelope is measured with a blower door test.

not exceed 5.0 air changes per hour or 0.28 cubic feet per minute (CFM) per square foot [0.0079 m³/(s × m²)] of dwelling unit enclosure area. Testing shall be conducted in accordance with ANSI/RESNET/ICC 380, ASTM E779 or ASTM E1827 and reported at a pressure of 0.2 inch w.g. (50 Pascals). Where required by the building official, testing shall be conducted by an approved third party. A written report of the results of the test shall be signed by the party conducting the test and provided to the building official. Testing shall be performed at any time after creation of all penetrations of the building thermal envelope have been sealed.

> **Exception:** For heated, attached private garages and heated, detached private garages accessory to one- and two-family dwellings and townhouses not more than three stories above grade plane in height, building envelope tightness and insulation installation shall be considered acceptable where the items in Table N1102.4.1.1, applicable to the method of construction, are field verified. Where required by the building official, an approved third party independent from the installer shall inspect both air barrier and insulation installation criteria. Heated, attached private garage space and heated, detached private garage space shall be thermally isolated from all other conditioned spaces in accordance with Sections N1102.2.12 and N1102.3.5, as applicable.

During testing:
Items 1 through 6 *(No changes)*

> **Exception:** When testing individual dwelling units, an air leakage rate not exceeding 0.30 cubic feet per minute per square foot [0.008 m³/(s × m²)] of the dwelling unit enclosure area tested in accordance with RESNET/ICC 380, ASTM E 779 or ASTM E 1827 and reported at a pressure of 0.2 inch w.g. (50 Pascals), shall be permitted in all climate zones for:
>
> 1. Attached single- and multiple-family building dwelling units.
> 2. Buildings or dwelling units that are 1,500 square feet (139.4 m²) or smaller.

Mechanical ventilation shall be provided in accordance with Section M1505 of this code or Section 403.3.2 of the *International Mechanical Code*, as applicable, or with other approved means of ventilation.

N1102.4.1.3 (R402.4.1.3) Leakage Rate. Where complying with Section N1101.13.1, the building or dwelling unit shall have an air leakage rate not exceeding 5.0 air changes per hour in Climate Zones 0, 1 and 2, and 3.0 air changes per hour in Climate Zones 3 through 8, when tested in accordance with Section N1102.4.1.2.

TABLE N1102.4.1.1 (R402.4.1.1) Air Barrier and Insulation Installation[a]

Component	Air Barrier Criteria	Insulation Installation Criteria
General requirements	A continuous air barrier shall be installed in the building envelope. ~~The exterior thermal envelope contains a continuous air barrier.~~ Breaks or joints in the air barrier shall be sealed.	Air-permeable insulation shall not be used as a sealing material.
Ceiling/attic	The air barrier in any dropped ceiling or soffit shall be aligned with the insulation and any gaps in the air barrier <u>shall be</u> sealed. Access openings, drop down stairs or knee wall doors to unconditioned attic spaces shall be sealed.	The insulation in any dropped ceiling/soffit shall be aligned with the air barrier.
Walls	The junction of the foundation and sill plate shall be sealed. The junction of the top plate and the top of exterior walls shall be sealed. Knee walls shall be sealed.	Cavities within corners and headers of frame walls shall be insulated by completely filling the cavity with a material having a thermal resistance, R-value, of not less than R-3 per inch. Exterior thermal envelope insulation for framed walls shall be installed in substantial contact and continuous alignment with the air barrier.
Windows, skylights and doors	The space between framing and skylights, and the jambs of windows and doors, shall be sealed.	—
Rim joists	Rim joists shall include ~~the~~ <u>an</u> exterior air barrier.[b] <u>The junctions of the rim board to the sill plate and the rim board toand the subfloor shall be air sealed.</u>	Rim joists shall be insulated <u>so that the insulation maintains permanent contact with the exterior rim board.</u>[b]
Floors including cantilevered floors and floors above garages.	The air barrier shall be installed at any exposed edge of insulation.	Floor framing cavity insulation shall be installed to maintain permanent contact with the underside of subfloor decking. Alternatively, floor framing cavity insulation shall be in contact with the top side of sheathing or continuous insulation installed on the underside of floor framing; and extending from the bottom to the top of all perimeter floor framing members.
<u>Basement,</u> ~~Crawl~~ crawl space ~~walls~~, <u>and slab foundations</u>	Exposed earth in unvented crawl spaces shall be covered with a Class I vapor retarder<u>/air barrier</u> ~~with overlapping joints taped~~ <u>in accordance with Section N1102.2.10. Penetrations through concrete foundation walls and slabs shall be air sealed. Class 1 Vapor retarders shall not be used as an air barrier on below-grade walls and shall be installed in accordance with Section R702.7.</u>	Crawl space insulation, where provided instead of floor insulation, shall ~~be permanently attached to the walls~~ <u>be installed in accordance with Section N1102.2.10. Conditioned basement foundation wall insulation shall be installed in accordance with Section N1102.2.8.1. Slab on grade floor insulation shall be installed in accordance with Section N1102.2.10.</u>
Shafts, penetrations	Duct <u>and flue</u> shafts, ~~utility penetrations, and flue shafts~~ <u>and other similar penetrations</u> ~~opening~~ to <u>the</u> exterior or unconditioned space shall be sealed <u>to allow for expansion, contraction, and mechanical vibration. Utility penetrations of the air barrier shall be caulked, gasketed or otherwise sealed and shall allow for expansion, contraction of materials and mechanical vibration.</u>	<u>Insulation shall be fitted tightly around utilities passing through shafts and penetrations in the building thermal envelope to maintain required R-value.</u>
Narrow cavities	<u>Narrow cavities of an inch or less that are not able to be insulated, shall be air sealed.</u>	Batts to be installed in narrow cavities shall be cut to fit or narrow cavities shall be filled with insulation that on installation readily conforms to the available cavity space.

(continues)

TABLE N1102.4.1.1 (R402.4.1.1) *(Continued)*

Component	Air Barrier Criteria	Insulation Installation Criteria
Garage separation	Air sealing shall be provided between the garage and conditioned spaces.	<u>Insulated portions of the garage separation assembly shall be installed in accordance with Sections R303 and N1102.2.7</u>
Recessed lighting	Recessed light fixtures installed in the building thermal envelope shall be <u>air</u> sealed ~~to the finished surface~~ in accordance with Section N1102.4.5.	Recessed light fixtures installed in the building thermal envelope shall be air tight and IC rated, and shall be buried or surrounded with insulation.
Plumbing, ~~and~~ wiring <u>or other obstructions</u>	<u>Holes created by wiring, plumbing or other obstructions in the air barrier assembly shall be air sealed.</u>	~~In exterior walls, batt insulation shall be cut neatly to fit around wiring and plumbing or insulation that on installation, readily conforms to available space, shall extend behind piping and wiring.~~ <u>Insulation shall be installed to fill the available space and surround wiring, plumbing, or other obstructions, unless the required *R*-value can be met by installing insulation and air barrier systems completely to the exterior side of the obstructions.</u>
Shower/tub on exterior wall	The air barrier installed at exterior walls adjacent to showers and tubs shall separate the wall from the shower or tub.	Exterior walls adjacent to showers, and tubs shall be insulated.
Electrical/phone box on exterior walls	The air barrier shall be installed behind electrical and communication boxes. Alternatively, air-sealed boxes shall be installed.	—
HVAC register boots	HVAC supply and return register boots that penetrate building thermal envelope shall be sealed to the subfloor, wall covering or ceiling penetrated by the boot.	—
Concealed sprinklers	Where required to be sealed, concealed fire sprinklers shall only be sealed in a manner that is recommended by the manufacturer. Caulking or other adhesive sealants shall not be used to fill voids between fire sprinkler cover plates and walls or ceilings.	—

a. Inspection of log walls shall be in accordance with the provisions of ICC 400.
b. <u>Air barrier and insulation full enclosure is not required in unconditioned/ventilated attic spaces and at rim joists.</u>

Note: Not all changes are shown. Please refer to the 2021 IRC for the complete text.

CHANGE SIGNIFICANCE: Several significant changes have been made to the building air leakage and testing provisions. For testing, the maximum air leakage rate has been set at five air changes per hour (ACH) for any compliance path. However, when following the prescriptive compliance path, the maximum leakage rate remains at 3 ACH for dwellings in Climate Zones 3 through 8. An exception was added for heated attached and detached garages that are field verified for air barrier and insulation installation criteria in the revised Table N1102.4.1.1. The ACH requirements for buildings and dwelling units following the prescriptive compliance path are relocated to new Section N1102.4.1.3.

An exception to using ACH to quantify air leakage in attached and small volume dwelling units was added because ACH is biased against small volume and attached dwellings. Although it is not difficult to get a single-family median size home to pass 3 or 5 ACH as required by the code, it is significantly more difficult to get a small volume home or an attached dwelling unit to pass. The alternative more accurately reflects leakage through the exterior enclosure area which removes built in volumetric bias while continuing to ensure a tight structure.

Changes also occur throughout Table N1102.4.1.1 Air Barrier and Insulation Installation. In many cases, the changes are editorial: redundant text has been removed and other provisions have been clarified. The component section of the table has been amended to include other obstructions besides wiring and plumbing piping. There are a number of obstructions in insulated building cavities that insulation must be split around so that it fully encloses the obstruction. For example, gas piping or HVAC duct work now can be included as obstructions. New references in the table point to specific sections of the code to clarify the details of installations. For example, recessed lighting fixture air leakage requirements are described in Section N1102.4.5.

N1102.4.6
Air-Sealed Electrical Boxes

CHANGE TYPE: Addition

CHANGE SUMMARY: Electrical and communication outlet boxes installed in the building thermal envelope must be sealed, tested and marked for compliance with the referenced standards.

2021 CODE: N1102.4.6 (R402.4.6) Electrical and communication outlet boxes (air-sealed boxes). Electrical and communication outlet boxes installed in the building thermal envelope shall be sealed to limit air leakage between conditioned and unconditioned spaces. Electrical and communication outlet boxes shall be tested in accordance with NEMA OS 4, *Requirements for Air-Sealed Boxes for Electrical and Communication Applications*, and shall have an air leakage rate of not greater than

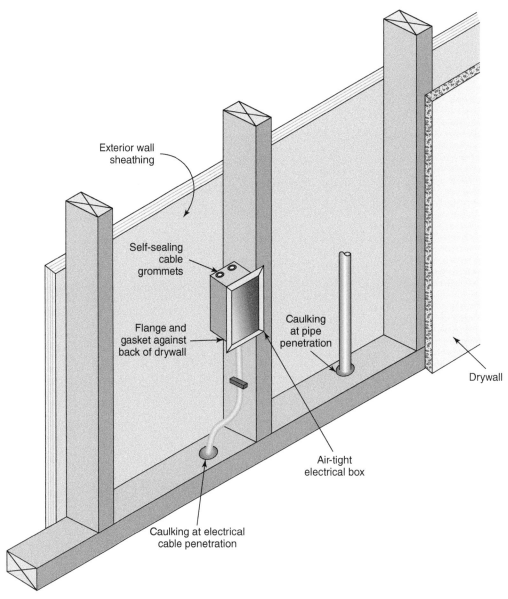

Electrical and communication outlet boxes in the thermal envelope must be sealed to limit air leakage.

2.0 cubic feet per minute (0.944 L/s) at a pressure differential of 1.57 psf (75 Pa). Electrical and communication outlet boxes shall be marked "NEMA OS 4" or "OS 4" in accordance with NEMA OS 4. Electrical and communication outlet boxes shall be installed per the manufacturer's instructions and with any supplied components required to achieve compliance with NEMA OS 4.

CHANGE SIGNIFICANCE: Similar to the requirements for recessed lighting, the new section sets specific air leakage requirements for electrical and communications outlet boxes to improve the efficiency of the building thermal envelope. The existing language in Table N1102.4.1.1 describes that the air barrier must be installed behind electrical and communication boxes, or alternatively, air-sealed boxes must be installed. The new Section N1102.4.6 defines air-sealed boxes as those meeting the prescribed standard and tested under the prescribed criteria. The new requirement will supersede other provisions in the code permitting onsite sealing of the air barrier, based on the concept that sealing air-barrier penetrations is not always as simple as applying more insulation, caulk, or expanding foam. Electrical and communication outlet boxes in the thermal envelope now must be tested and marked in compliance with the air-sealing standards and installed in accordance with the manufacturer's instructions.

N1103.3
Duct Installation

CHANGE TYPE: Modification

CHANGE SUMMARY: Duct insulation requirements have been clarified in Sections N1103.3.1 through N1103.3.3.1

2021 CODE: N1101.6 (R202). Thermal Distribution Efficiency (TDE). The resistance to changes in air heat as air is conveyed through a distance of air duct. TDE is a heat loss calculation evaluating the difference in the heat of the air between the air duct inlet and outlet caused by differences in temperatures between the air in the duct and the duct material. TDE is expressed as a percent difference between the inlet and outlet heat in the duct.

N1103.3 (R403.3) Ducts. Ducts and air handlers shall be installed in accordance with Sections N1103.3.1 through N1103.3.7.

N1103.3.1 (R403.3.1) Ducts ~~Insulation~~ located outside conditioned space ~~(Prescriptive)~~. Supply and return ducts ~~in attics~~ located outside conditioned space shall be insulated to an *R*-value of not less than R-8 for ducts 3 inches (76 mm) in diameter and larger and not less than R-6 for ducts smaller than 3 inches (76 mm) in diameter. ~~Supply and return ducts in other portions of the building shall be insulated to not less than R-6 for ducts 3 inches (76 mm) in diameter and to not less than R-4.2 for ducts smaller than 3 inches (76.2 mm) in diameter.~~ Ducts buried beneath a building shall be insulated as required by this section or have an equivalent thermal distribution efficiency. Underground ducts utilizing the thermal distribution efficiency method shall be listed and labeled to indicate the *R*-value equivalency.

> **Exception:** ~~Ducts or portions thereof located completely inside the building thermal envelope.~~

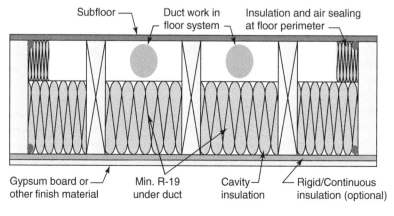

Ductwork considered to be in conditioned space.

~~N1103.3.7 (R403.3.7)~~ <u>N1103.3.2 (R403.3.2)</u> **Ducts located in conditioned space.** For ~~ducts~~ <u>ductwork</u> to be considered ~~as~~ inside ~~a~~ conditioned space, ~~such ducts~~ <u>it</u> shall comply with ~~either~~ <u>one</u> of the following:

1. The duct system is located completely within the continuous air barrier and within the building thermal envelope.
2. ~~The ducts are~~ <u>Ductwork in ventilated attic spaces is</u> buried within ceiling insulation in accordance with Section ~~N1103.3.6~~ <u>N1103.3.3</u> and all of the following conditions exist:
   2.1. The air handler is located completely within the continuous air barrier and within the building thermal envelope.
   2.2. The duct leakage, as measured either by a rough-in test of the ducts or a post-construction total system leakage test to outside the building thermal envelope in accordance with Section ~~N1103.3.4~~ <u>N1103.3.6</u>, is less than or equal to 1.5 cubic feet per minute (42.5 L/min) per 100 square feet (9.29 m^2) of conditioned floor area served by the duct system.
   2.3. The ceiling insulation *R*-value installed against and above the insulated duct is greater than or equal to the proposed ceiling insulation *R*-value, less the *R*-value of the insulation on the duct.
3. <u>Duct work in floor cavities located over unconditioned space shall have the following:</u>
   3.1. <u>A continuous air barrier installed between unconditioned space and the duct</u>
   3.2 <u>Insulation installed in accordance with Section N1102.2.7</u>
   3.3. <u>A minimum *R*-19 insulation installed in the cavity width separating the duct from unconditioned space.</u>
4. <u>Duct work located within exterior walls of the building thermal envelope shall have the following:</u>
   4.1. <u>A continuous air barrier installed between unconditioned space and the duct</u>
   4.2. <u>Minimum R-10 insulation installed in the cavity width separating the duct from the outside sheathing</u>
   4.3 <u>The remainder of the cavity insulation fully insulated to the drywall side.</u>

~~N1103.3.6 (R403.3.6)~~ <u>N1103.3.3 (R403.3.3)</u> **Ducts buried within ceiling insulation.**

~~N1103.3.6.1 (R403.3.6.1)~~ <u>N1103.3.3.1 (R403.3.3.1)</u> **Effective *R*-value of deeply buried ducts.**

~~N1103.3.2 (R403.3.2)~~ <u>N1103.3.4 (R403.3.4)</u> **Sealing (~~Mandatory~~).**

~~N1103.3.2.1 (R403.3.2.1)~~ <u>N1103.3.4.1 (R403.3.4.1)</u> **Sealed air handler.**

~~N1103.3.3 (R403.3.3)~~ <u>N1103.3.5 (R403.3.5)</u> **Duct testing (~~Mandatory~~).**

~~N1103.3.4 (R403.3.4)~~ <u>N1103.3.6 (R403.3.6)</u> **Duct leakage (~~Prescriptive~~).**

~~N1103.3.5 (R403.3.5)~~ <u>N1103.3.7 (R403.3.7)</u> Building cavities (~~Mandatory~~).

Note: Not all changes are shown. Please refer the 2021 IRC for the complete text.

CHANGE SIGNIFICANCE: In the 2018 code, the addition of provisions for ducts buried within ceiling insulation addressed the installation issue for ducts located outside of the continuous air barrier assemblies. The new language completes this transition by addressing ducts that are located within the continuous air barrier and the building thermal envelope assemblies in concealed locations. Section N1103.3 now clearly defines the four scenarios where ductwork can be installed and qualify as being in conditioned space:

- Duct system completely within the continuous air barrier and building thermal envelope
- Duct work in ventilated attic and buried within ceiling insulation
- Duct work in floor cavities over unconditioned space
- Duct work within exterior walls

The new language addresses minimum requirements to separate the ductwork from unconditioned spaces with prescribed insulation and air barrier, including installations in floor assemblies and walls. Keeping the ducts inside conditioned space increases the energy performance of homes.

The entire Section N1103.3 was reorganized for clarity and application:

- N1103.3.1 Ducts located outside conditioned space
- N1103.3.2 Ducts located in conditioned space
- N1103.3.3 Ducts buried within ceiling insulation
- N1103.3.4 Duct Sealing
- N1103.3.5 Duct Testing
- N1103.3.6 Duct Leakage
- N1103.3.7 Building Cavities

N1103.3.5
Duct Testing

CHANGE TYPE: Modification

CHANGE SUMMARY: The exception for duct testing of ducts located entirely in conditioned spaces has been removed.

2021 CODE: ~~**N1103.3.3 (R403.3.3)**~~ <u>**N1103.3.5 (R403.3.5)**</u> Duct testing ~~**(Mandatory)**~~. Ducts shall be pressure tested <u>in accordance with ANSI/RESNET/ICC 380 or ASTM E1554</u> to determine air leakage by one of the following methods:

1. Rough-in test: Total leakage shall be measured with a pressure differential of 0.1 inch w.g. (25 Pa) across the system, including the manufacturer's air handler enclosure if installed at the time of the test. Registers shall be taped or otherwise sealed during the test.

2. Postconstruction test: Total leakage shall be measured with a pressure differential of 0.1 inch w.g. (25 Pa) across the entire system, including the manufacturer's air handler enclosure. Registers shall be taped or otherwise sealed during the test.

<u>**Exceptions:**</u>

~~1. A duct air-leakage test shall not be required where the ducts and air handlers are located entirely within the building thermal envelope.~~

~~2.~~ A duct air-leakage test shall not be required for ducts serving ~~heat or energy recovery ventilators~~ <u>ventilation systems</u> that are not integrated with ducts serving heating or cooling systems.

A written report of the results of the test shall be signed by the party conducting the test and provided to the building official.

Ductwork entirely within the building thermal envelope requires testing.

~~N1103.3.4 (R403.3.4)~~ **N1103.3.6 (R403.3.6) Duct leakage ~~(Prescriptive)~~.** The total leakage of the ducts, where measured in accordance with Section N1103.3.5, shall be as follows:

1. Rough-in test: The total leakage shall be less than or equal to 4.0 cubic feet per minute (113.3 L/min) per 100 square feet (9.29 m²) of conditioned floor area where the air handler is installed at the time of the test. Where the air handler is not installed at the time of the test, the total leakage shall be less than or equal to 3.0 cubic feet per minute (85 L/min) per 100 square feet (9.29 m²) of conditioned floor area.

2. Postconstruction test: Total leakage shall be less than or equal to 4.0 cubic feet per minute (113.3 L/min) per 100 square feet (9.29 m²) of conditioned floor area.

3. <u>Tests for ducts within thermal envelope: Where all ducts and air handlers are located entirely within the building thermal envelope, total air leakage shall be less than or equal to 8.0 cubic feet per minute (226.6 L/min) per 100 square feet (9.29 m²) of conditioned floor area.</u>

CHANGE SIGNIFICANCE: The code previously exempted homes from duct testing requirements where the air handler and all ducts were located inside conditioned space. Although moving all ducts inside conditioned space may have a positive impact on energy efficiency overall, this practice alone cannot guarantee that the ducts will be tight enough to deliver conditioned air to all occupied areas of the home. Uncomfortable occupants commonly adjust thermostat settings to counteract the effect of poor delivery of conditioned air, leading to losses in energy efficiency.

This code change requires a duct leakage test for all new dwellings, including those with all ducts inside conditioned space. Note that the allowable leakage rate will be set at twice the prescriptive rate when all ducts are located inside conditioned space.

Duct leakage rates can be extremely high when ducts are not tested. Without an objective test as a means of quality assurance, even careful builders may not be aware of missed connections or poor sealing. In a recent Department of Energy (DOE) field study of residential homes in Kentucky, homes received duct leakage tests even where all supply and return ducts were located inside conditioned space. Of the 24 homes tested with ducts inside conditioned spaces, the leakage rate averaged 18.5 cfm/sqft, with individual homes ranging from 6.26 cfm/sqft to as high as 40.36 cfm/sqft, higher leakage rates than the 2018 IECC requirement. Forty other homes in the same study with at least some ducts located outside conditioned space achieved leakage rates of 9.7 cfm/sqft, on average – roughly half the leakage rate of homes that qualified for the exemption. The field studies found similar results in Pennsylvania, where "exempt" homes (with all ducts inside conditioned space) averaged 31 cfm/sqft leakage, while homes required to be tested averaged 18 cfm/sqft leakage.

N1103.6
Mechanical Ventilation

CHANGE TYPE: Addition

CHANGE SUMMARY: A heat or energy recovery ventilation system is required for Climate Zones 7 and 8. Mechanical ventilation systems now require testing.

2021 CODE: N1103.6 (R403.6) Mechanical ventilation (Mandatory). ~~The building~~ Buildings and dwelling units shall be provided with mechanical ventilation that complies with the requirements of Section M1505 or with other approved means of ventilation. Outdoor air intakes and exhausts shall have automatic or gravity dampers that close when the ventilation system is not operating.

N1103.6.1 (R403.6.1) Heat or energy recovery ventilation. Dwelling units shall be provided with a heat recovery or energy recovery ventilation system in Climate Zones 7 and 8. The system shall be balanced with a minimum sensible heat recovery efficiency of 65 percent at 32°F (0°C) at a flow greater than or equal to the design airflow.

~~N1103.6.1 (R403.6.1)~~ N1103.6.2 (R403.6.2) Whole-~~house~~ dwelling mechanical ventilation system fan efficacy. Fans used to provide whole-~~house~~ dwelling mechanical ventilation shall meet the efficacy requirements of Table ~~N1103.6.1~~ N1103.6.2 at one or more rating points. Fans shall be tested in accordance with HVI 916 and listed. The airflow

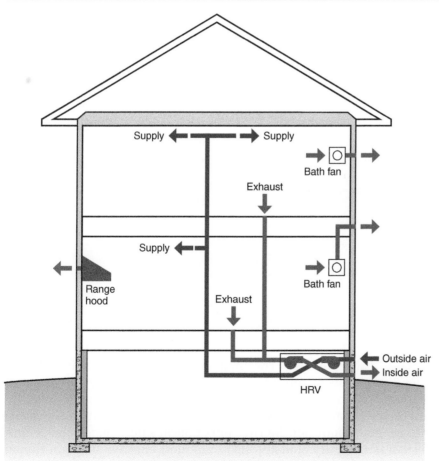

An HRV or ERV is required in Climate Zones 7 and 8.

shall be reported in the product listing or on the label. Fan efficacy shall be reported in the product listing or shall be derived from the input power and airflow values reported in the product listing or on the label. Fan efficacy for fully ducted HRV, ERC, balanced, and in-line fans shall be determined at a static pressure of not less than 0.2 inch w.c. (49.82 Pa). Fan efficacy for ducted range hoods, bathroom, and utility room fans shall be determined at a static pressure of not less than 0.1 inch w.c. (24.91 Pa).

> **Exception:** ~~Where an air handler that is integral to tested and listed HVAC equipment is used to provide whole-house mechanical ventilation, the air handler shall be powered by an electronically commutated motor.~~

N1103.6.3 (R403.6.3) Testing. Mechanical ventilation systems shall be tested and verified to provide the minimum ventilation flow rates required by Section N1103.6. Testing shall be performed according to the ventilation equipment manufacturer's instructions, or by using a flow hood or box, flow grid, or other airflow measuring device at the mechanical ventilation fan's inlet terminals or grilles, outlet terminals or grilles, or in the connected ventilation ducts. Where required by the building official, testing shall be conducted by an approved third party. A written report of the results of the test shall be signed by the party conducting the test and provided to the building official.

> **Exception:** Kitchen range hoods that are ducted to the outside with 6 inch (152 mm) or larger duct and not more than one 90-degree elbow or equivalent in the duct run.

TABLE ~~N1103.6.1 (R403.6.1)~~ N1103.6.2 (R403.6.2)   Whole-~~House~~ Dwelling Mechanical Ventilation System Fan Efficacy[a]

System Type ~~Fan Location~~	Air Flow Rate Minimum (cfm)	Minimum Efficacy (cfm/watt)[a]	~~Air Flow Rate Maximum (CFM)~~
HRV, ~~or~~ ERV or balanced	Any	1.2 cfm/watt	~~Any~~
~~Range hoods~~	~~Any~~	~~2.8 cfm/watt~~	~~Any~~
In-line supply or exhaust fan	Any	~~2.8~~ 3.8 cfm/watt	~~Any~~
Other exhaust fan ~~Bathroom, utility room~~	~~10~~ < 90	~~1.4~~ 2.8 cfm/watt	~~< 90~~
Other exhaust fan ~~Bathroom, utility room~~	≥ 90	~~2.8~~ 3.5 cfm/watt	~~Any~~
Air-handler that is integrated to tested and listed HVAC equipment	Any	1.2 cfm/watt	

For SI: 1 cubic foot per minute = 28.3 L/min.

~~a. When tested in accordance with HVI Standard 916.~~
a. Design outdoor airflow rate/watts of fan used.

CHANGE SIGNIFICANCE: Dwelling units located in the coldest Climate Zones 7 and 8 now require a heat recovery (HRV) or energy recovery ventilation (ERV) system to save energy, particularly in the colder months. With these systems, heat loss is minimized by using conditioned exhaust air to warm incoming air. An HRV only transfers heat energy and is most effective in cold, dry climates. ERVs are most effective in warm humid climates.

The energy provisions require mechanical ventilation and the minimum ventilation rates are indicated in Section N1103.6. The testing experience gained through the verification of the EnergyStar program has demonstrated that ventilation fans are installed but are not always performing as required by the code. Fan rated flow does not necessarily equate to the flow that is actually produced once a fan has been installed. The quality of the installation of the duct from the fan to the termination of the duct to the outside as well as the quality of the termination device ultimately governs the amount of air that any fan can push. This new provision requires testing to ensure that the systems are not only there but have also been installed in such a way that they work as intended by the code.

This new provision applies to all compliance paths and is included in Table N1105.2 for the total building performance path and Table N1106.2 for the ERI compliance path. Requiring testing of spot (local) and whole house ventilation systems intends to move the building industry into compliance with the code by offering direct feedback on the fan choice and the installation. This feedback will guide fan choice and installation techniques that will be in compliance with the code.

N1104

Lighting Equipment

CHANGE TYPE: Addition

CHANGE SUMMARY: High-efficacy lighting is now required in all permanent lighting fixtures. New provisions regulate lighting controls for interior and exterior lighting.

2021 CODE: N1104.1 (R404.1) Lighting equipment (Mandatory). ~~Not less than 90 percent of the~~ <u>All</u> permanently installed lighting fixtures<u>, excluding kitchen appliance lighting fixtures,</u> shall contain only high-efficacy <u>lighting sources</u> ~~lamps~~.

<u>**N1104.2 (R404.2) Interior Lighting Controls.** Permanently installed lighting fixtures shall be controlled with a dimmer, an occupant sensor control or another control that is installed or built into the fixture.</u>

<u>**Exception:** Lighting controls shall not be required for the following:</u>
<u>1. Bathrooms</u>
<u>2. Hallways</u>
<u>3. Exterior lighting fixtures</u>
<u>4. Lighting designed for safety or security</u>

<u>**N1104.3 (R404.3) Exterior lighting controls.** Where the total permanently installed exterior lighting power is greater than 30 watts, the permanently installed exterior lighting shall comply with the following:</u>

<u>1. Lighting shall be controlled by a manual on and off switch that permits automatic shut off actions.</u>

<u>**Exception:** Lighting serving multiple dwelling units.</u>

<u>2. Lighting shall be automatically shut off when daylight is present and satisfies the lighting needs.</u>

<u>3. Controls that override automatic shut off actions shall not be allowed unless the override automatically returns automatic control to its normal operation within 24 hours.</u>

Controls are required for exterior lighting exceeding 30 watts.

All permanently installed lighting fixtures require high-efficacy light sources.

CHANGE SIGNIFICANCE: Section N1104 broadly addresses residential electrical power and lighting systems. Lighting uses about 10 percent of residential electrical power. New to the 2021 IRC electrical power and lighting provisions are grouped in the following categories:

- N1104.1 High-efficacy lighting sources for all permanent lighting fixtures
- N1104.2 Interior lighting controls
- N1104.3 Exterior lighting controls

Historically the IECC, and therefore Chapter 11 of the IRC, has not included specific requirements for residential lighting controls. However, as seen by its impact on the commercial market, controls applied to equipment and lighting can provide further energy efficiencies. The new Section N1104.2 requires interior lighting controls in the form of either a dimmer, occupant sensor, or some other control. Both dimmers and occupancy controls save energy. Dimmers can reduce energy use by about 20 percent, while occupant sensors reduce wasted energy by around 30 percent. These controls are essentially permanent, or at least have a long lifetime. This new provision applies to all compliance paths and is included in Table N1105.2, Requirements for Total Building Performance and Table N1106.2, Requirements for Energy Rating Index.

The new Section N1104.3 introduces residential exterior lighting provisions to the code. The new requirement includes controls for systems with greater than 30 watts – manual switches that allow automatic shutoff, automatic shutoff based on daylight, and overrides for the controls. The type of control is not specified, only what it must accomplish. The 2014 Consortium for Energy Efficiency (CEE) report "Lighting Controls Market Characterization Report," identified that the use of a photosensor or timer can save, on average, up to 60 KWh per year, based on the efficacy of the light source that is controlled. This new provision for exterior lighting applies only to the prescriptive compliance path.

N1105 and Table N1105.2

Total Building Performance Analysis

CHANGE TYPE: Modification

CHANGE SUMMARY: The Total Building Performance compliance path has been retitled and reorganized with a new Table N1105.2 listing the related mandatory requirements that appear elsewhere in the code.

2021 CODE:

SECTION N1105 (R405)
~~SIMULATED PERFORMANCE ALTERNATIVE (PERFORMANCE)~~
TOTAL BUILDING PERFORMANCE

N1105.1 (R405.1) Scope. This section establishes criteria for compliance using ~~simulated energy~~ <u>total building</u> performance analysis. Such analysis shall include heating, cooling, mechanical ventilation and service water heating energy only.

N1105.2 (R405.2) ~~Mandatory requirements.~~ <u>Performance based compliance.</u> Compliance ~~with this section~~ <u>based on total building performance</u> requires that ~~the mandatory provisions identified in Section N1101.13 be met.~~ <u>a proposed design meets all of the following:</u> ~~Supply and return ducts not completely inside the building thermal envelope shall be insulated to an R-value of not less than R-6.~~

<u>1. The requirements of the sections indicated within Table N1105.2</u>

<u>2. The building thermal envelope greater than or equal to levels of efficiency and solar heat gain coefficients in Table R402.1.1 or R402.1.3 of the 2009 *International Energy Conservation Code*.</u>

<u>3. An annual energy cost that is less than or equal to the annual energy cost of the standard reference design. Energy prices shall be taken from a source approved by the building official, such as the Department of Energy, Energy Information Administration's State</u>

Fenestration orientation and shading strategies are part of a Total Building Performance evaluation.

Energy Data System Prices and Expenditures reports. The building official shall be permitted to require time-of-use pricing in energy cost calculations.

Exception: The energy use based on source energy expressed in Btu or Btu per square foot of conditioned floor area shall be permitted to be substituted for the energy cost. The source energy multiplier for electricity shall be 3.16. The source energy multiplier for fuels other than electricity shall be 1.1.

TABLE N1105.2 (R405.2) Requirements for Total Building Performance

Section[a]	Title
General	
N1101.13.5	Additional energy efficiency
N1101.14	Certificate
Building Thermal Envelope	
N1102.1.1	Vapor Retarder
N1102.2.3	Eave Baffles
N1102.2.4.1	Access hatches and doors
N1102.2.10.1	Crawl space wall insulation installation
N1102.4.1.1	Installation
N1102.4.1.2	Testing
N1102.5	Maximum fenestration U-factor and SHGC
Mechanical	
N1103.1	Controls
N1103.3, including N1103.3.1.1, except Sections N1103.3.2, N1103.3.3, and N1103.3.6	Ducts
N1103.4	Mechanical system piping insulation
N1103.5.1	Heated water circulation and temperature maintenance systems
N1103.5.3	Drain water heat recovery units
N1103.6	Mechanical ventilation
N1103.7	Equipment sizing and efficiency rating
N1103.8	Systems serving multiple dwelling units
N1103.9	Snow melt system controls
N1103.10	Energy consumption of pools and permanent spas
N1103.11	Portable spas
N1103.12	Residential pools and permanent residential spas
Electrical Power and Lighting Systems	
N1104.1	Lighting equipment
N1104.2	Interior Lighting Controls

a. Reference to a code section includes all the relative subsections except as indicated in the table.

N1105.3 (R405.3) Performance-based compliance. ~~Compliance based on simulated energy performance requires that a proposed residence (proposed design) be shown to have an annual energy cost that is less than or equal to the annual energy cost of the standard reference design. Energy prices shall be taken from a source approved by the code official, such as the Department of Energy, Energy Information Administration's State Energy Data System Prices and Expenditures reports. Code officials shall be permitted to require time-of-use pricing in energy cost calculations.~~

> **Exception:** ~~The energy use based on source energy expressed in Btu or Btu per square foot of conditioned floor area shall be permitted to be substituted for the energy cost. The source energy multiplier for electricity shall be 3.16. The source energy multiplier for fuels other than electricity shall be 1.1.~~

Note: Not all changes to Section N1105 are shown. Please refer to the 2021 IRC for the complete text.

CHANGE SIGNIFICANCE: Section N1105 contains the provisions for the Total Building Performance compliance option. This simulated performance alternative offers flexibility to determine the most cost-effective methods to design a code-compliant structure. This method uses software to predict the annual energy use of the proposed design compared to that of the standard design of building components listed in Table N1105.4.2(1), Specifications for the Standard Reference and Proposed Designs. If the proposed building's annual energy use is less than or equal to that of the standard design, the building complies with the code's goal of effective energy use. This method is useful to consider for any building designs that incorporate shading strategies to minimize air-conditioning loads in the summer and maximize heat gain in the winter.

The 2021 IECC and Chapter 11 of the IRC include restructuring of the section to clarify required provisions (formerly referred to as mandatory) and updates to the standard reference design.

The labels "prescriptive" and "mandatory" were introduced into the 2009 IRC. The terms have not been defined and have not been used consistently. These terms were applied to various section and subsection titles throughout Chapter 11 creating confusion for users. Furthermore, the labels were not requirements and were not enforceable.

Among long-time code users, it is generally understood that "mandatory" was intended to mean "nontradeable" when using performance compliance options, meaning that where the procedures or systems described within the "mandatory" section were included as part of the design, the requirements of that section had to be met and could not be traded off. "Prescriptive," on the other hand, was intended to mean "mandatory" when using the prescriptive path, but "tradeable" when using the performance path. However, this understanding was not explicitly based on code language and new users to the code have been confused about the use of the terms.

The 2021 IRC has eliminated the terms mandatory and prescriptive and instead all requirements applicable to the Total Building Performance compliance path are identified within Section N1105. Table N1105.2 lists the section number and title of all required provisions for this compliance path.

Since the 2009 code, the Total Building Performance alternative has seen extensive use in Colorado and that experience continues to inform the language of the IECC and IRC Chapter 11. In the 2015 edition, the process by which the Simulated Performance path was successfully implemented in Colorado was included as language in the body of the code to help jurisdictions and builders better understand how to implement the pathway to gain flexibility of choice and trade-offs.

In the 2021 code the Colorado experience further informs the code, with an update of the compliance requirements. The changes provide clarification and simplification rather than actually changing requirements. For example, the requirements for the submittal to obtain a building permit and steps to achieve a certificate of occupancy are more clearly laid out. In addition, Section N1105.4.2.1 Compliance Report for Permit Application and Section N1105.4.2.2 Confirmed Compliance Report for Certificate of Occupancy, have been consolidated and reordered to more clearly identify the documentation needed to be reviewed to demonstrate compliance.

N1106 and Table N1106.2

Energy Rating Index Analysis

CHANGE TYPE: Modification

CHANGE SUMMARY: New Table N1106.2 lists the requirements associated with the Energy Rating Index (ERI) compliance path. ERI values have been lowered to improve energy efficiency.

2021 CODE:

SECTION N1106 (R406)
ENERGY RATING INDEX COMPLIANCE ALTERNATIVE

N1106.1 (R406.1) Scope. This section establishes criteria for compliance using an Energy Rating Index (ERI) analysis.

N1106.2 (R406.2) Mandatory requirements. Compliance with this section requires that the provisions identified in Sections N1101 through N1104 indicated as "Mandatory" and Section N1103.5.3 be met. The building thermal envelope shall be greater than or equal to levels of efficiency and Solar Heat Gain Coefficients in Table N1102.1.1 or N1102.1.3 of the 2009 International Energy Conservation Code.

The Energy Rating Index is determined in accordance with RESNET/ICC 301.

~~**Exception:** Supply and return ducts not completely inside the building thermal envelope shall be insulated to an R-value of not less than R-6.~~

N1106.2 (R406.2) ERI Compliance. Compliance based on the Energy Rating Index (ERI) requires that the rated design meets all of the following:

1. The requirements of the sections indicated within Table N1106.2
2. Maximum energy rating index of Table N1106.5

TABLE N1106.2 (R406.2) Requirements for Energy Rating Index

Section[a]	Title
General	
N1101.13.5	Additional Efficiency Packages
N1101.14	Certificate
Building Thermal Envelope	
N1102.1.1	Vapor Retarder
N1102.2.3	Eave Baffles
N1102.2.4	Access hatches and doors
N1102.2.10.1	Crawl space wall insulation installations
N1102.4.1.1	Installation
N1102.4.1.2	Testing
Mechanical	
N1103.1	Controls
N1103.3 except sections N1103.3.2, N1103.3.3, and N1103.3.6	Ducts
N1103.4	Mechanical system piping insulation
N1103.5.3	Heated water circulation and temperature maintenance systems
N1103.5.3	Drain water heat recovery units
N1103.6	Mechanical ventilation
N1103.7	Equipment sizing and efficiency rating
N1103.8	Systems serving multiple dwelling units
N1103.9	Snow melt system controls
N1103.10	Energy consumption of pools and spas
N1103.11	Portable spas
N1103.12	Residential pools and permanent residential spas
Electrical Power and Lighting Systems	
N1104.1	Lighting equipment
N1104.2	Interior Lighting Controls
N1106.3	Building Thermal Envelope

a. Reference to a code section includes all of the relative subsections except as indicated in the table.

N1106.3 Building Thermal Envelope.

N1106.3.1 On-site renewables are not included.

N1106.3.2 On-site renewables are included.

N1106.4 ~~N1106.3~~ **Energy Rating Index.**

N1106.5 ~~N1106.4~~ **ERI-based compliance.**

TABLE ~~N1106.4 (R406.4)~~ **N1106.5 (R406.5)** Maximum Energy Rating Index

Climate Zone	Energy Rating Index[a]
0-1	~~57~~ 52
2	~~57~~ 52
3	~~57~~ 51
4	~~62~~ 54
5	~~61~~ 55
6	~~61~~ 54
7	~~58~~ 53
8	~~58~~ 53

a. ~~Where on-site renewable energy is included for compliance using the ERI analysis of Section N1106.4, the building shall meet the mandatory requirements of Section N1106.2, and the building thermal envelope shall be greater than or equal to the levels of efficiency and SHGC in Table N1102.1.2 or Table N1102.1.4.~~

N1106.6 ~~N1106.5~~ **Verification by approved agency.**

N1106.7 ~~N1106.6~~ **Documentation.**

N1106.7.1 ~~N1106.6.1~~ **Compliance software tools.**

N1106.7.2 ~~N1106.6.2~~ **Compliance report.**

N1106.7.2.1 ~~N1106.6.2.1~~ **Proposed compliance report for permit application.**

Note: Due to space limitations, not all changes for Section N1106 are shown. Please refer to the 2021 IRC for the complete text.

CHANGE SIGNIFICANCE: Section N1106 contains the provisions for the ERI compliance path. Introduced into the 2015 code, it is another path that can be used to demonstrate compliance with the energy code. The ERI is an index scoring system with a number based on the ERI reference design value of 100. A residential building that uses no net purchased energy has an index value of 0. The ERI reference design is configured such that it meets the minimum requirements of the 2006 IECC. Each number on the scale represents a 1 percent change in the total energy use of the rated design relative to the total energy use of the ERI reference design benchmarked at 100 on the scale. The ERI score considers all energy used in the residential building.

The 2021 code includes restructuring of the section to clarify required provisions (formerly referred to as mandatory), updates the required ERI scores, and introduces provisions for renewable energy certificate documentation.

The labels "prescriptive" and "mandatory" were applied to various section and subsection titles throughout the code creating confusion for users. Furthermore, the labels were not requirements and were not enforceable.

The 2021 code has eliminated the terms mandatory and prescriptive and instead all requirements applicable to the N1106 Energy Rating Index are identified within Section N1106. New Table N1106.2 lists the section number and title of all required provisions for this compliance path.

Table N1106.5 is revised to restore the lower ERI target scores that appeared in the 2015 IECC and 2015 IRC. Coupled with the new requirements in Section N1101.13.5, Additional Energy Efficiency, the ERI compliance option will result in a significantly more efficient building than was required under the 2018 code. The additional energy efficiency measures require buildings complying under the energy rating index to use an ERI value that is at least 5 percent less than the ERI target specified in Table N1106.5 (already lowered in the 2021 code), making the ERI approach significantly more stringent than the other compliance paths.

N1108

Additional Efficiency Package Options

CHANGE TYPE: Addition

CHANGE SUMMARY: The new requirement for choosing an energy efficiency package option, in addition to the prescriptive and total building performance compliance path options, will increase energy efficiency of new dwellings.

2021 CODE:

SECTION N1108 (R408)
ADDITIONAL EFFICIENCY PACKAGE OPTIONS

N1108.1 (R408.1) Scope. This section establishes additional efficiency package options to achieve additional energy efficiency in accordance with Section N1101.13.5.

N1108.2 (R408.2) Additional efficiency package options. Additional efficiency package options for compliance with Section N1101.13.5 are set forth in Sections N1108.2.1 through N1108.2.5.

N1108.2.1 (R408.2.1) Enhanced envelope performance option The total building thermal envelope UA, the sum of U-factor times assembly area, shall be less than or equal to 95 percent of the total UA resulting from multiplying the U-factors in Table N1102.1.2 by the same assembly area as in the proposed building. The UA calculation shall be performed in accordance with Section N1102.1.5. The area-weighted average SHGC of all glazed fenestration shall be less than or equal to 95 percent of the maximum glazed fenestration SHGC in Table N1102.1.2.

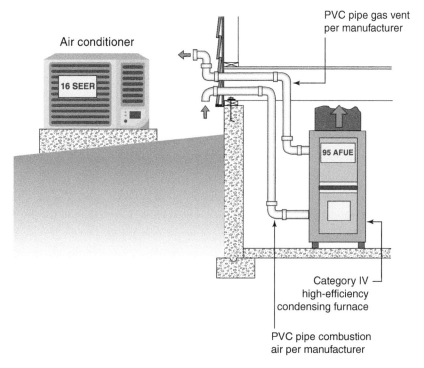

16 SEER AC and 95 AFUE furnace additional efficiency option.

N1108.2.2 (R408.2.2) More efficient HVAC equipment performance option. Heating and cooling equipment shall meet one of the following efficiencies:

1. Greater than or equal to 95 AFUE natural gas furnace and 16 SEER air conditioner.
2. Greater than or equal to 10 HSPF/16 SEER air source heat pump.
3. Greater than or equal to 3.5 COP ground source heat pump. For multiple cooling systems, all systems shall meet or exceed the minimum efficiency requirements in this section and shall be sized to serve 100 percent of the cooling design load. For multiple heating systems, all systems shall meet or exceed the minimum efficiency requirements in this section and shall be sized to serve 100 percent of the heating design load.

N1108.2.3 (R408.2.3) Reduced energy use in service water-heating option. The hot water system shall meet one of the following efficiencies:

1. Greater than or equal to 82 EF fossil fuel service water-heating system.
2. Greater than or equal to 2.0 EF electric service water-heating system.
3. Greater than or equal to 0.4 solar fraction solar water-heating system.

N1108.2.4 (R408.2.4) More efficient duct thermal distribution system option. The thermal distribution system shall meet one of the following efficiencies:

1. 100 percent of ducts and air handlers located entirely within the building thermal envelope.
2. 100 percent of ductless thermal distribution system or hydronic thermal distribution system located completely inside the building thermal envelope.
3. 100 percent of duct thermal distribution system located in conditioned space as defined by N1103.3.2.

N1108.2.5 (R408.2.5) Improved air sealing and efficient ventilation system option. The measured air leakage rate shall be less than or equal to 3.0 ACH50, with either an energy recovery ventilator (ERV) or heat recovery ventilator (HRV) installed. Minimum HRV and ERV requirements, measured at the lowest tested net supply airflow, shall be greater than or equal to 75 percent sensible recovery efficiency (SRE), less than or equal to 1.1 cubic feet per minute per watt (0.03 m^3/min/watt) and shall not use recirculation as a defrost strategy. In addition, the ERV shall be greater than or equal to 50 percent latent recovery/moisture transfer (LRMT).

CHANGE SIGNIFICANCE: Arguably the most significant change to Chapter 11 of the IRC and the residential provisions of the 2021 IECC, new Section N1108 (IECC Section R408) coupled with scoping in Section N1101.13.5, represents additional requirements applied to the prescriptive and total building performance compliance paths.

The application of Section N1108 is established in N1101.13.5:

- Prescriptive compliance projects must include one package from Section N1108.
- Total building performance compliance projects must either include one package of additional requirements in the project, without including it in the proposed design calculations, or the proposed design must have an annual energy cost less than or equal to 95 percent of the standard reference design.

Note: Projects complying via the ERI path do not require any of the additional efficiency package options in Section N1108, but must have a score at least 5 percent less than that in the ERI Table N1106.4.

The new Section N1108 largely mirrors the format of the 2012 through 2018 editions of the commercial provisions of IECC Section C406, Additional Efficiency Package Options. Like Section C406, new IRC Section N1108 with companion Section N1101.13.5 offers multiple improvements that will increase energy savings and reduce energy costs to the homeowner over the useful life of the building.

Section N1108 is modeled after code language adopted in states like Oregon and Washington that have successfully created a list of options available to builders to comply with improvements to the base residential code requirements. This approach is intended to allow flexibility for code users while advancing a building's energy efficiency.

PART 5
Mechanical
Chapters 12 through 23

- **Chapter 12** Mechanical Administration
  No changes addressed
- **Chapter 13** General Mechanical System Requirements
  No changes addressed
- **Chapter 14** Heating and Cooling Equipment
  No changes addressed
- **Chapter 15** Exhaust Systems
- **Chapter 16** Duct Systems
  No changes addressed
- **Chapter 17** Combustion Air
  No changes addressed
- **Chapter 18** Chimneys and Vents
- **Chapter 19** Special Appliances, Equipment and Systems
  No changes addressed
- **Chapter 20** Boilers and Water Heaters
  No changes addressed
- **Chapter 21** Hydronic Piping
- **Chapter 22** Special Piping and Storage Systems No changes addressed
- **Chapter 23** Solar Thermal Energy Systems
  No changes addressed

As a comprehensive code that applies to all aspects of residential construction, the IRC contains provisions for the mechanical, fuel gas, plumbing and electrical systems of the building. These systems are covered in their respective parts of the IRC beginning with Part 5. Chapter 12 contains administrative provisions unique to the application and enforcement of regulations governing mechanical systems, as well as the technical provisions related to system design and installation in Chapters 13 through 23. Chapter 13 provides the general requirements for all mechanical systems and addresses the listing and labeling of appliances, types of fuel used, access to appliances, clearance to combustibles and other related issues. The remainder of Part 5 deals with requirements for specific mechanical systems related to heating and cooling, exhaust, ventilation, ducts, vents, boilers and hydronic piping. The last two chapters of Part 5 contain provisions specific to fuel oil piping and storage, and thermal solar energy systems. ■

M1505
Balanced Ventilation System Credit

M1802.4
Blocked Vent Switch for Oil-fired Appliances

M2101
Hydronic Piping Systems Installation

223

M1505
Balanced Ventilation System Credit

CHANGE TYPE: Modification

CHANGE SUMMARY: The code now allows a 30 percent reduction to the mechanical ventilation airflow rate for balanced ventilation systems.

2021 CODE: R202 Definitions.

<u>**BALANCED VENTILATION SYSTEM.** A ventilation system where the total supply airflow and total exhaust airflow are simultaneously within 10 percent of their average. The balanced ventilation system airflow is the average of the supply and exhaust airflows.</u>

M1505.1 General. Where local exhaust or whole-house mechanical ventilation is provided, the ~~equipment~~ <u>ventilation system</u> shall be designed in accordance with this section.

M1505.3 Exhaust equipment. Exhaust ~~equipment serving single dwelling units~~ <u>fans and whole-house mechanical ventilation fans</u> shall be listed and labeled as providing the minimum required airflow in accordance with ANSI/AMCA 210-ANSI/ASHRAE 51.

M1505.4.2 System controls. The whole-house mechanical ventilation system shall be provided with <u>controls that enable manual override. Controls shall include text or a symbol indicating their function.</u>

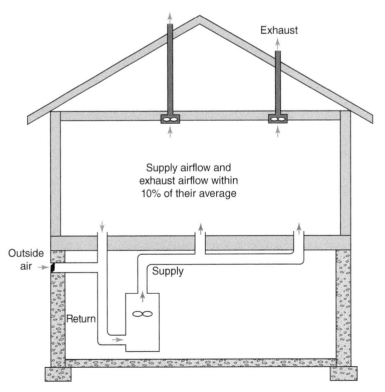

Balanced ventilation system.

M1505.4.3 Mechanical ventilation rate. The whole house mechanical ventilation system shall provide outdoor air at a continuous rate ~~as~~ not less than that determined in accordance with Table M1505.4.3(1) or not less than that determined by Equation 15-1.

[Equation 15-1]

Ventilation rate in cubic feet per minute = (0.01 x total square foot area of house) + [7.5 x (number of bedrooms + 1)]

Exceptions:

1. Ventilation rate credit. The minimum mechanical ventilation rate determined in accordance with Table M1505.4.3(1) or Equation 15-1 shall be reduced by 30%, provided that both of the following conditions apply:
   1.1. A ducted system supplies ventilation air directly to each bedroom and to one or more of the following rooms:
      1.1.1. Living room
      1.1.2. Dinning room
      1.1.3. Kitchen
   1.2. The whole-house ventilation system is a balanced ventilation system.
2. Programmed intermittent operation. The whole-house mechanical ventilation system is permitted to operate intermittently where the system has controls that enable operation for not less than 25 percent of each 4-hour segment and the ventilation rate prescribed in Table M1505.4.3(1) , by Equation 15-1, or by Exception 1 is multiplied by the factor determined in accordance with Table M1505.4.3(2).

TABLE M1505.4.4 Minimum Required Local Exhaust Rates for One- and Two-Family Dwellings

Area to be exhausted	Exhaust rates[a]
Kitchens	100 cfm intermittent or 25 cfm continuous
Bathrooms—Toilet Rooms	Mechanical exhaust capacity of 50 cfm intermittent or 20 cfm continuous

For SI: 1 cubic foot per minute = 0.0004719 m³/s. 1 inch water column = 0.2488 kPa

a. The listed exhaust rate for bathrooms-toilet rooms shall equal or exceed the exhaust rate at a minimum static pressure of 0.25 inch wc in accordance with Section M1505.3.

CHANGE SIGNIFICANCE: Balanced mechanical ventilation systems provide superior ventilation to unbalanced systems. The new provisions for reduced air flow in balanced systems recognize the advantages of balanced systems and they are no longer required to provide the same airflow rate as less effective, unbalanced systems to provide equivalent ventilation. The credit to reduce the required airflow rate for balanced ventilation is based on a simplified version derived from ASHRAE 62.2-2016 Equation 4.2. The ASHRAE equation adjusts the balanced whole house ventilation flow rate as a function of building air leakage, building height, and weather and shielding factor (which approximates a corresponding climate zone). To simplify application of the ASHRAE calculation, the code recognizes a one-size-fits-all balanced system factor based on a methodology developed for a typical new, single-family detached home with one-, two- and three-stories. The assigned air leakage rates are conservative values assuming that the average home is slightly tighter than the maximum allowed leakage rates of the energy efficiency provisions.

To ensure that exhaust fans provide the minimum CFM required by the IRC, the 2015 edition added Table M1504.2 to provide prescriptive duct sizing based on fan exhaust rates taken at a static pressure of 0.25-inch water column. For consistency, footnote a has been added to Table 1505.4.4 to align the flow rate requirements with the duct sizing requirements of Table M1504.2 and the equipment listing requirements of Section M1505.3.

M1802.4 Blocked Vent Switch for Oil-fired Appliances

CHANGE TYPE: Addition

CHANGE SUMMARY: An additional safety device for oil-fired appliances has been added to be consistent with what is required for some gas-fired appliances.

2021 CODE: M1802.4 Blocked vent switch. Oil-fired appliances shall be equipped with a device that will stop burner operation in the event that the venting system is obstructed. Such device shall have a manual reset and shall be installed in accordance with the manufacturer's instructions.

CHANGE SIGNIFICANCE: Blocked (obstructed) vent switches are not new, but the code has not required them until now. Oil-fired appliances such as furnaces vent to masonry chimneys, factory-built chimneys or Type L vents. These venting systems, especially masonry chimneys, are subject to blockage by plant debris, animal nesting materials, bird and raccoon carcasses, deteriorated masonry and similar objects. A blocked vent switch detects spillage from the venting system, typically from a draft regulator installed between the appliance and the vent or chimney and shuts off the appliance. This function can prevent carbon monoxide poisoning in the event of an obstructed vent or chimney. The manual reset feature prevents the appliance from operating without intentional action by an occupant or an HVAC service person. If the blocked vent switch activates, the cause must be determined and remedied before the appliance can resume operation.

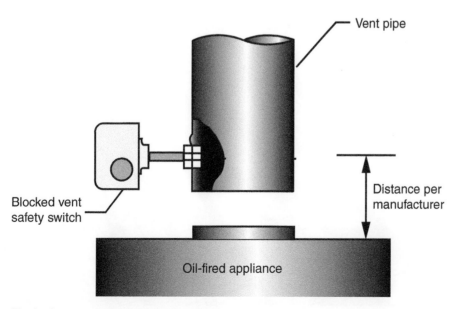

Blocked vent switch for oil-fired appliances.

M2101

Hydronic Piping Systems Installation

CHANGE TYPE: Modification

CHANGE SUMMARY: The provisions for ground source heat pump loop piping systems in Section M2105 have been duplicated in Section M2101 to apply to all hydronic piping systems.

2021 CODE:

Section M2101
Hydronic Piping Systems Installation

M2101.11 Used materials. Used pipe, fittings, valves, and other materials shall not be reused in hydronic systems.

M2101.12 Material rating. Pipe and tubing shall be rated for the operating temperature and pressure of the system. Fittings shall be suitable for the pressure applications and recommended by the manufacturer for installation with the pipe and tubing material installed. Where used underground, materials shall be suitable for burial.

M2101.13 Joints and connections. Joints and connections shall be of an approved type. Joints and connections shall be tight for the pressure of the system. Joints used underground shall be approved for such applications.

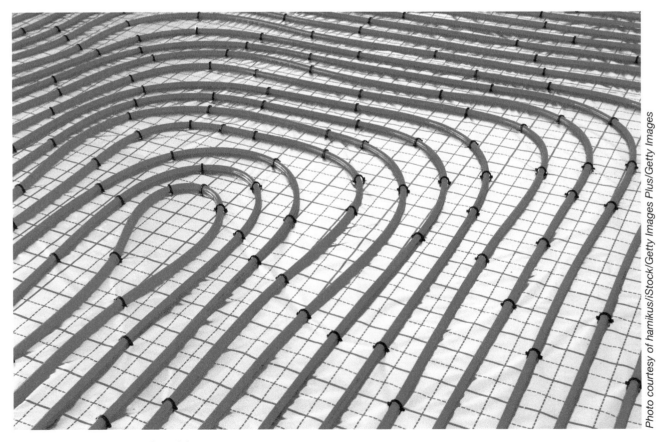

Hydronic radiant floor heating piping.

M2101.13.1 Joints between different piping materials. Joints between different piping materials shall be made with approved transition fittings.

M2101.14 Preparation of pipe ends. Pipe shall be cut square and shall be free of burrs and obstructions. Pipe ends shall have full-bore openings and shall be prepared in accordance with the pipe manufacturer's instructions.

M2101.15 Joint preparation and installation. Where required by Sections M2101.16 through M2101.18, the preparation and installation of mechanical and thermoplastic-welded joints shall comply with Sections M2101.15.1 and M2101.15.2.

M2101.15.1 Mechanical joints. Mechanical joints shall be installed in accordance with the manufacturer's instructions.

M2101.15.2 Thermoplastic-welded joints. Joint surfaces for thermoplastic-welded joints shall be cleaned by an approved procedure. Joints shall be welded in accordance with the manufacturer's instructions.

M2101.16 CPVC plastic pipe. Joints between CPVC plastic pipe or fittings shall be solvent-cemented in accordance with Section P2906.9.1.2. Threaded joints between fittings and CPVC plastic pipe shall be in accordance with Section M2101.16.1

M2101.16.1 Threaded joints. Threads shall conform to ASME B1.20.1. The pipe shall be Schedule 80, 40 or heavier plastic pipe and shall be threaded with dies specifically designed for plastic pipe. Thread lubricant, pipe-joint compound or tape shall be applied on the male threads only and shall be approved for application on the piping material.

M2101.17 Cross-linked polyethylene (PEX) plastic tubing. Joints between cross-linked polyethylene plastic tubing and fittings shall comply with Sections M2101.17.1 and M2101.17.2. Mechanical joints shall comply with Section M2101.15.1.

M2101.17.1 Compression-type fittings. Where compression-type fittings include inserts and ferrules or O-rings, the fittings shall be installed without omitting the inserts and ferrules or O-rings.

Section M2103
Floor Heating Systems

M2103.1 Piping materials. Piping for embedment in concrete or gypsum materials shall be standard-weight steel pipe, copper and copper-alloy pipe and tubing, cross-linked polyethylene/aluminum/cross-linked polyethylene (PEX-ALPEX) pressure pipe, chlorinated polyvinyl chloride (CPVC), ~~polybutylene,~~ cross-linked polyethylene (PEX) tubing, polyethylene of raised temperature (PE-RT) or polypropylene (PP) with a rating of not less than ~~100~~ 80 pounds per square inch at 180°F (~~690~~ 552 kPa at 82°C).

Section M2105
Ground-Source Heat-Pump System Loop Piping

M2105.7 Preparation of pipe ends. Pipe shall be cut square, ~~reamed,~~ and shall be free of burrs and obstructions. ~~CPVC, PE and PVC pipe shall be chamfered.~~ Pipe ends shall have full-bore openings and shall ~~not be undercut~~ be prepared in accordance with the pipe manufacturer's instructions.

Note: Due to space limitations, not all changes are shown for Chapter 21. Please refer to the 2021 IRC for the complete text.

CHANGE SIGNIFICANCE: The general hydronic systems provisions have been expanded and updated to provide more detailed guidance for installations. Section M2101 now provides the same level of coverage as is currently provided for ground-source heat pump systems in Section M2105. Sections M2101.1 through M2101.10 remain without significant changes and cover basics such as piping and fitting standards, piping support and testing. The added Sections M2101.11 through M2101.30 are pulled directly from Section M2105 and primarily cover piping and fitting materials, joint and connection methods for the various materials, valves and pressure vessels and various other components and safeguards.

A change to Section M2103.1 Piping materials for floor heating systems, recognizes that piping for embedment in concrete or gypsum materials does not need to be rated at 100 psi. Hydronic heating systems are typically designed with operating pressures of 12 psi–20 psi, and these systems contain expansion tanks incorporated in them that are factory set to 12 psi. Safety relief valves on the boilers are typically set at 30 psi or 50 psi. The code now permits piping rated at 80 psi and is consistent with information in ASTM F2623, *Standard Specification for Polyethylene of Raised Temperature (PE-RT) SDR 9 Tubing*.

In Section M2105.7 dealing with preparation of pipe ends for ground source heat pump system loop piping, the text has been revised to accurately reflect procedures for the type of piping being used. This section is specific to plastic pipes and some of the existing language referred to terms such as "reamed" and "undercut" which only apply to metallic pipes. These terms have been deleted. The requirement for CPVC, PE and PVC pipe ends to be chamfered has also been removed. The revised language is more appropriate by requiring the pipe ends to be prepared in accordance with the manufacturer's instructions. The instructions will be specific to the type of plastic pipe being used.

PART 6
Fuel Gas
Chapter 24

- **Chapter 24** Fuel Gas

Fuel gas systems are covered in Part 6, including provisions for approved materials as well as the design and installation of fuel gas piping and other system components. The fuel gas provisions of the IRC are taken directly from the *International Fuel Gas Code* (IFGC). In order to make the correlation and coordination of the two codes easier, after each fuel gas section of the IRC the corresponding section of the IFGC is shown in parentheses. The fuel gas portion of the IRC contains its own specific definitions in Section G2403 in addition to the general definitions found in Chapter 2 of the IRC. The text, tables and figures in other sections of Chapter 24 address the technical issues of fuel gas systems, such as appliance installation; materials, sizing, and installation of fuel gas piping systems; piping support; flow controls; connections; combustion air; venting; and other related system requirements. ■

G2403
Definitions of Point of Delivery and Service Meter Assembly

G2414.8.3
Threaded Joint Sealing

G2415.5
Fittings in Concealed Locations

G2427.5.5.1
Chimney Lining

G2427.8
Through-the-wall Vent Terminal Clearances

G2439.5
Makeup Air for Dryer Installed in a Closet

G2447.2
Commercial Cooking Appliances Prohibited

G2403
Definitions of Point of Delivery and Service Meter Assembly

CHANGE TYPE: Modification

CHANGE SUMMARY: Revisions to definitions clarify the portions of the gas piping system regulated by the serving utility and the portions regulated by the IRC fuel gas provisions downstream of the point of delivery.

2021 CODE:

Section G2403 (202)
General Definitions

Point of Delivery. For natural gas systems, the point of delivery is the outlet of the service meter assembly or the outlet of the service regulator or service shutoff valve where a meter is not provided. Where a <u>system shutoff</u> valve is provided ~~at~~ <u>after</u> the outlet of the service meter assembly, such valve shall be considered to be downstream of the point of delivery. For undiluted liquefied petroleum gas systems, the point of delivery shall be considered to be the outlet of the service pressure regulator, exclusive of line gas regulators, in the system.

<u>**Service Meter Assembly.** The meter, valve, regulator, piping, fittings and equipment installed by the service gas supplier before the point of delivery.</u>

<u>**System Shutoff.** A valve installed after the point of delivery to shut off the entire piping system.</u>

Valve. A device used in piping to control the gas supply to any section of a system of piping or to an appliance.

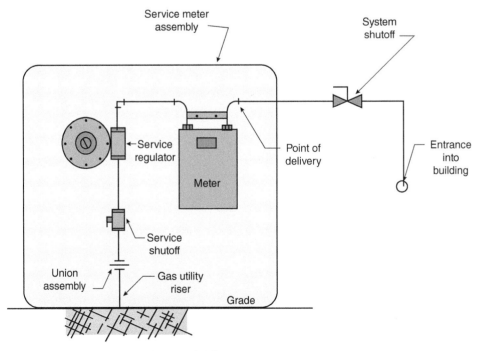

Service meter assembly and point of delivery.

Service Shutoff. A valve, installed by the serving gas supplier between the ~~service meter or~~ source of supply and the ~~customer piping system~~ <u>point of delivery</u>, to shut off the entire piping system.

CHANGE SIGNIFICANCE: The new and revised definitions intend to clarify the transition from the gas utility company service to the portion of gas piping system regulated by the IRC.

The "point of delivery" (POD) is the point where the fuel gas provisions of the code first take effect. Downstream of the POD is governed by the code and upstream of the POD is governed by the Department of Transportation (DOT) and the gas utility company. The definition of POD refers to the "service meter assembly," but this latter term has not previously been defined. The new definition clarifies what components are under the control of the gas utility company and are, therefore, upstream of the POD and not controlled by IRC provisions. The definition of "service shutoff" (valve) was revised to refer to the POD instead of the vague reference to "customer piping system." The new definition of "service meter assembly" includes the meter as one of several components that are upstream of the POD.

The definition of POD was also revised to include the newly defined term "system shutoff," which is an optional valve installed downstream of the service meter assembly. A system shutoff valve is installed by choice to allow the customer or building owner to shut off the gas to the entire building. This valve also allows the gas to be turned on by a party other than the gas utility, after the gas utility has set the service meter assembly.

The distinction between a service shutoff valve and a system shutoff valve is that the latter is owned and operated by the customer and the service shutoff valve is owned by the gas utility and can be operated only by qualified personnel working for the gas utility.

G2414.8.3
Threaded Joint Sealing

CHANGE TYPE: Modification

CHANGE SUMMARY: Thread joint sealants are now required for assembling threaded joints in gas piping.

2021 CODE: ~~G2414.9.3 (403.9.3)~~ <u>G2414.8.3 (403.8.3)</u> **Thread<u>ed</u> joint ~~compounds~~ <u>sealing</u>.** <u>Threaded joints shall be made using a thread joint sealing material.</u> Thread joint <u>sealing materials</u> ~~compounds~~ shall be <u>nonhardening and shall be</u> resistant to the ~~action of liquefied petroleum gas or to any other~~ chemical constituents of the gases to be ~~conducted~~ <u>conveyed</u> through the piping. <u>Thread joint sealing materials shall be compatible with the pipe and fitting materials on which the sealing materials are used.</u>

CHANGE SIGNIFICANCE: The code previously addressed pipe thread compounds, stating that when used, they would resist any chemical constituents of the gases. Until now, the code has not required the use of thread sealants on threaded joints. Thread sealants act primarily as a lubricant to allow the threads to make up tight to form a metal-to-metal seal. Any imperfections or voids in the threads are filled in by the thread sealant material. The most common thread sealants used as of this writing are pastes made with PTFE (Teflon) and Teflon tapes. Sealing compounds must be of material that does not harden over time so it remains effective and so that pipe joints can be disassembled when necessary. Shrinkage of the sealant is minimized so that the material remains resilient and resistant to vibration and movement. The code maintains the requirement that the sealant must resist the effects of any components of the gas being conveyed in the pipe. The reference to liquefied petroleum gas (LPG) has been removed because it was redundant. The reference to the chemical constituents of all gases to be conducted includes LPG.

Threaded joint sealing tape.

Photo courtesy of Vudhikul Ocharoen/iStock / Getty Images Plus/Getty Images

The thread sealant used must also be suitable for the pipe and fitting materials used. Although this is typically not an issue, some thread sealants can attack certain types of pipe and fitting materials. This is more of an issue with plastic threads used in plumbing piping, whereas threaded fittings for gas piping are always metal. Thread sealant container labels will state what pipe and fitting materials are suitable for use with the sealant.

G2415.5

Fittings in Concealed Locations

CHANGE TYPE: Clarification

CHANGE SUMMARY: Plugs and caps have been added to the list of threaded fittings approved for concealed locations.

2021 CODE: G2415.5 (404.5) Fittings in concealed locations. Fittings installed in concealed locations shall be limited to the following types:

1. Threaded elbows, tees, ~~and~~ couplings, plugs and caps.
2. Brazed fittings.
3. Welded fittings.
4. Fittings listed to ANSI LC1/CSA 6.26 or ANSI LC4/CSA 6.32.

CHANGE SIGNIFICANCE: Threaded fittings, such as those used with steel gas piping, are allowed to be concealed, with the exception of unions. The list of approved fittings for concealment appear in Item 1 of Section G2415.5. The omission of threaded plugs and threaded caps was an oversight and did not intend to imply that these fittings could not be concealed. To correct the omission and to provide a complete list of approved fittings, plugs and caps have been added to the threaded fittings list. Although left/right couplings are sometimes viewed as a type of union, they are threaded couplings and are not prohibited.

Threaded fittings permitted in concealed locations.

Significant Changes to the IRC 2021 Edition G2427.5.5.1 ■ Chimney Lining

G2427.5.5.1
Chimney Lining

CHANGE TYPE: Modification

CHANGE SUMMARY: The exception allowing an existing chimney to vent replacement appliances has been deleted.

2021 CODE: G2427.5.5.1 (503.5.6.1) Chimney lining. Chimneys shall be lined in accordance with NFPA 211.

> **Exception:** Where an existing chimney complies with Sections G2427.5.5 through G2427.5.5.3 and its sizing is in accordance with Section G2427.5.4, its continued use shall be allowed where the appliance vented by such chimney is replaced by an appliance of similar type, input rating and efficiency.

CHANGE SIGNIFICANCE: The exception to the chimney lining provisions previously allowed an existing unlined chimney to continue to be used when replacing the appliance vented by the chimney. The exception stipulated that the replacement appliance had to be similar in type, Btu/h input rating and thermal efficiency. The code has otherwise required a replacement appliance to comply with the provisions for a new appliance installation including venting.

The exception applied to all appliances. For example, a replacement water heater installation is required to be served by a code-compliant venting system, just as the initial water heater was. However, this exception allowed a noncode-compliant chimney to serve a new water heater installation, in conflict with a basic principle of the code, the appliance installation instructions and the appliance listing. The original purpose of the exception was to remove a potential financial hardship caused by requiring an old unlined existing chimney to be lined when the water

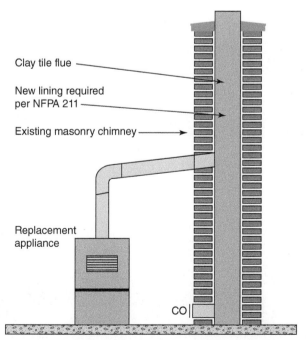

A chimney liner meeting NFPA 211 is required for replacement appliances.

heater had to be replaced. The exception was based on the assumption that the previous water heater was venting properly and the existing unlined chimney was in serviceable condition. Because newer appliances change design, input ratings and efficiency, it is unlikely that a replacement appliance will be similar to the previous appliance. Serviceable unlined chimneys are also unusual. The exception was considered outdated and unnecessary, and has been deleted from the code.

Significant Changes to the IRC 2021 Edition G2427.8 ■ Through-the-wall Vent Terminal Clearances **239**

G2427.8
Through-the-wall Vent Terminal Clearances

CHANGE TYPE: Clarification

CHANGE SUMMARY: Through-the-wall vent terminal clearance distances have been placed in a new table with a corresponding figure for ease of use.

2021 CODE: G2427.8 (503.8) Venting system ~~termination location~~ terminal clearances. The clearances for through-the-wall direct-vent and nondirect-vent terminals shall be in accordance with Table G2427.8 and Figure G2427.8.

Exception: The clearances in Table G2427.8 shall not apply to the combustion air intake of a direct-vent appliance

~~The location of venting system terminations shall comply with the following (see Appendix C):~~

~~1. A mechanical draft venting system shall terminate not less than 3 feet (914 mm) above any forced-air inlet located within 10 feet (3048 mm).~~

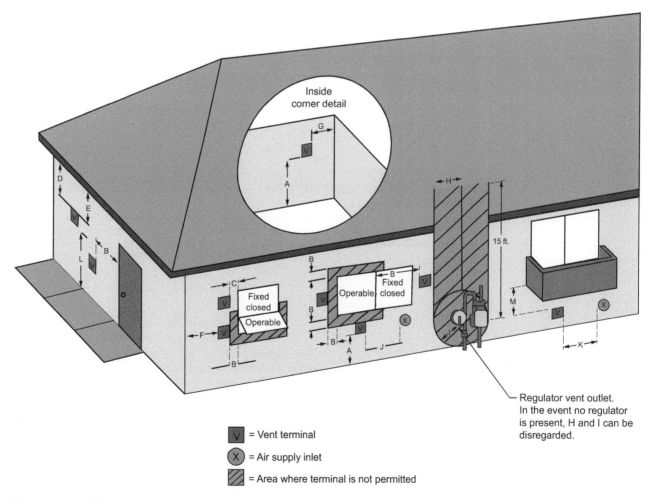

Through-the-wall vent terminal clearances.

Exceptions:

1. ~~This provision shall not apply to the combustion air intake of a direct-vent appliance.~~
2. ~~This provision shall not apply to the separation of the integral outdoor air inlet and flue gas discharge of listed outdoor appliances.~~

2. ~~A mechanical draft venting system, excluding direct-vent appliances, shall terminate not less than 4 feet (1219 mm) below, 4 feet (1219 mm) horizontally from, or 1 foot (305 mm) above any door, operable window or gravity air inlet into any building. The bottom of the vent terminal shall be located not less than 12 inches (305 mm) above finished ground level.~~
3. ~~The clearances for through-the-wall, direct-vent terminals shall be in accordance with Table G2427.8.~~
4. ~~Through-the-wall vents for Category II and IV appliances and noncategorized condensing appliances shall not terminate over public walkways or over an area where condensate or vapor could create a nuisance or hazard or could be detrimental to the operation of regulators, relief valves or other equipment. Where local experience indicates that condensate is a problem with Category I and III appliances, this provision shall also apply. Drains for condensate shall be installed in accordance with the appliance and vent manufacturer's instructions.~~
5. ~~Vent systems for Category IV appliances that terminate through an outside wall of a building and discharge flue gases perpendicular to the adjacent wall shall be located not less than 10 feet (3048 mm) horizontally from an operable opening in an adjacent building. This requirement shall not apply to vent terminals that are 2 feet (607 mm) or more above or 25 feet (7620 mm) or more below operable openings.~~

~~**TABLE G2427.8 (503.8)** Through-The-Wall, Direct-Vent Termination Clearances~~

~~Direct-Vent Appliance Input Rating (Btu/hr)~~	~~Through-The-Wall Vent Terminal Clearance from Any Air Opening into the Building (inches)~~
~~< 10,000~~	~~6~~
~~≥ 10,000 ≤ 50,000~~	~~9~~
~~> 50,000 ≤ 150,000~~	~~12~~
~~> 150,000~~	~~In accordance with the appliance manufacturer's instructions and not less than the clearances specified in Section G2427.8, Item 2~~

~~For SI: 1 inch = 25.4 mm, 1 Btu/hr = 0.2931 W.~~

TABLE G2427.8 (503.8) Through-the-Wall Vent Terminal Clearances

Figure Clearance	Clearance Location	Minimum clearances for Direct-Vent Terminals	Minimum clearances for Non-Direct Vent Terminals
A	Clearance above finished grade level, veranda, porch, deck, or balcony	12 inches	
B	Clearance to window or door that is openable	6 inches: Appliances ≤ 10,000 Btu/hr 9 inches: Appliances > 10,000 Btu/hr ≤ 50,000 Btu/hr 12 inches: Appliances > 50,000 Btu/hr ≤ 150,000 Btu/hr Appliances > 150,000 Btu/hr, in accordance with the appliance manufacturer's instructions and not less than the clearances specified for non-direct vent terminals in Row B	4 feet below or to side of opening or 1 foot above opening
C	Clearance to non-openable window	None unless otherwise specified by the appliance manufacturer	
D	Vertical clearance to ventilated soffit located above the terminal within a horizontal distance of 2 feet from the center line of the terminal	None unless otherwise specified by the appliance manufacturer	
E	Clearance to unventilated soffit	None unless otherwise specified by the appliance manufacturer	
F	Clearance to outside corner of building	None unless otherwise specified by the appliance manufacturer	
G	Clearance to inside corner of building	None unless otherwise specified by the appliance manufacturer	
H	Clearance to each side of center line extended above regulator vent outlet	3 feet up to a height of 15 feet above the regulator vent outlet	
I	Clearance to service regulator vent outlet in all directions	3 feet for gas pressures up to 2 psi 10 feet for gas pressures above 2 psi	
J	Clearance to non-mechanical air supply inlet to building and the combustion air inlet to any other appliance	Same clearance as specified for row B	
K	Clearance to a mechanical air supply inlet	10 feet horizontally from inlet or 3 feet above inlet	
L	Clearance above paved sidewalk or paved driveway located on public property	7 feet and shall not be located above public walkways or other areas where condensate or vapor can cause a nuisance or hazard	
M	Clearance to underside of veranda, porch deck, or balcony	12 inches where the area beneath the veranda, porch deck or balcony is open on not less than two sides. The vent terminal is prohibited in this location where only one side is open.	

For SI units, 1 inch = 25.4 mm, 1 foot = 304.8 mm, 1 pound per square inch = 6.895 kPa, 1 Btu/hr = 0.293 W

CHANGE SIGNIFICANCE: The reformatting of Section G2427.8 is not meant to make any technical changes to the through-the-wall vent terminal clearance provisions. Reorganizing the content and placing all of the clearance requirements into a single table makes the information easier to locate and understand than the previous layout of the material. In addition, new Figure G2427.8 illustrates the location and clearances for the vent terminations. The identifying letter in the table corresponds to the lettered locations in the figure. The figure is similar to the figure that has appeared in Appendix C of the IRC.

G2439.5

Makeup Air for Dryer Installed in a Closet

CHANGE TYPE: Clarification

CHANGE SUMMARY: The requirement for a transfer opening for supplying makeup air to a closet designed for a gas dryer has been moved into a separate section.

2021 CODE: G2439.5 (614.7) Makeup air. Installations exhausting more than 200 cfm (0.09 m³/s) shall be provided with makeup air. ~~Where a closet is designed for the installation of a clothes dryer, an opening having an area of not less than 100 square inches (645 mm2) for makeup air shall be provided in the closet enclosure, or makeup air shall be provided by other approved means.~~

<u>**G2439.5.1 (614.7.1) Closet Installation.** Where a closet is designed for the installation of a clothes dryer, an opening having an area of not less than 100 square inches (645 mm²) for makeup air shall be provided in the closet enclosure, or makeup air shall be provided by other approved means.</u>

CHANGE SIGNIFICANCE: Section G2439.5 requires all dryers exhausting greater than 200 cfm to be provided with a makeup air supply because of the concern for creating negative pressure in the space and starving the dryer for airflow. This requirement is separate from the requirement for providing an opening in closet enclosures that contain dryers. Including both in one section has caused some confusion. With both requirements in one section in the 2018 edition, the code appeared to require the closet opening only for dryers that exhaust over 200 cfm. Section G2439.5 now addresses only the requirement to provide makeup air for dryers that exceed the 200 cfm threshold and Section G2439.5.1 addresses the requirement to have an opening in the closet enclosure to allow makeup air into the closet. Without the opening into the closet, the dryer would be isolated from its source of makeup air and the closet would experience significant negative pressure. The requirement for closets in Section G2439.5.1 applies to all dryers, regardless of the exhaust rate.

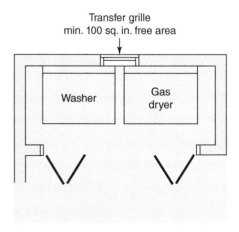

Makeup air required for gas dryer installed in a closet.

G2447.2
Commercial Cooking Appliances Prohibited

CHANGE TYPE: Modification

CHANGE SUMMARY: The exception allowing a commercial cooking appliance in a dwelling unit when the installation is designed by an engineer has been removed from the code.

2021 CODE: G2447.2 (623.2) Prohibited location. Cooking appliances designed, tested, listed and labeled for use in commercial occupancies shall not be installed within dwelling units or within any area where domestic cooking operations occur.

Exceptions:
1. ~~1.~~ Appliances that are also listed as domestic cooking appliances.
2. ~~Where the installation is designed by a licensed Professional Engineer, in compliance with the manufacturer's installation instructions.~~

CHANGE SIGNIFICANCE: Commercial cooking appliances are manufactured to different standards and do not incorporate the same safety features as domestic cooking appliances. For example, commercial ranges operate at higher temperatures, have higher input ratings, have less insulation to protect the user, require greater clearances to combustibles, may require a noncombustible floor and lack child-safe features. Adequate ventilation and makeup air also become a concern with the installation of commercial cooking appliances with high velocity range hoods. For those reasons, the fuel gas provisions have historically prohibited installation of commercial cooking appliances within dwelling units. The 2015 IRC fuel gas provisions introduced an exception to allow appliances that were listed for both commercial and domestic use. Trends in recent years have seen an increase in large homes with the latest in kitchen appliances

Cooking appliances must be listed for domestic use.

installed in very large kitchens. In turn there has been an increased demand for commercial-type cooking appliances in residences. Some states and local jurisdictions have recognized this by amending their ordinances to allow commercial appliances under certain conditions. In a new exception in the 2018 IRC, installation of commercial cooking appliances within dwelling units was specifically permitted by the code provided the installation was designed by a licensed professional engineer and the design complied with the manufacturer's instructions. That exception has been removed from the 2021 code.

The marketplace has taken care of the demand for commercial-style appliances. Appliance manufacturers currently offer many commercial-style appliances that are dual listed as both commercial and household appliances. This dual listing is already addressed in the first exception.

PART 7
Plumbing
Chapters 25 through 33

- Chapter 25   Plumbing Administration
- Chapter 26   General Plumbing Requirements
  No changes addressed
- Chapter 27   Plumbing Fixtures
- Chapter 28   Water Heaters
  No changes addressed
- Chapter 29   Water Supply and Distribution
- Chapter 30   Sanitary Drainage
- Chapter 31   Vents
  No changes addressed
- Chapter 32   Traps
  No changes addressed
- Chapter 33   Storm Drainage
  No changes addressed

Part 7 of the IRC contains provisions for plumbing systems and begins with a chapter on the specific and unique administrative issues related to plumbing code enforcement. Subsequent chapters cover technical subjects for the overall design and installation of plumbing systems in buildings. General plumbing issues such as protection of plumbing systems from damage, piping support, and certification of products are covered in Chapter 26.

The other chapters of Part 7 are specific to requirements for plumbing fixtures, water heaters, water supply and distribution, sanitary drainage, vents, traps, and storm drainage. ■

P2503.5.1
Drain, Waste and Vent Systems Testing

P2708.4, P2713.3
Shower and Bathtub Control Valves

P2904
Installation Practices for Residential Sprinklers

P2905.3
Length of Hot Water Piping to Fixtures

P3005.2.10.1
Removable Fixture Traps as Cleanouts

P3011
Relining of Building Sewers and Building Drains

P2503.5.1
Drain, Waste and Vent Systems Testing

CHANGE TYPE: Modification

CHANGE SUMMARY: The head pressure for a water test of drain, waste and vent (DWV) systems has increased from 5 feet to 10 feet. Air vacuum testing is now permitted for plastic piping DWV systems.

2021 CODE: P2503.5.1 Rough plumbing. DWV systems shall be tested on completion of the rough piping installation by water ~~or~~, <u>by air for piping systems other than plastic,</u> ~~by air,~~ <u>or by a vacuum of air for plastic piping systems,</u> without evidence of leakage. ~~Either~~ <u>The</u> test shall be applied to the drainage system in its entirety or in sections after rough-in piping has been installed, as follows:

1. Water test. Each section shall be filled with water to a point not less than ~~5~~ <u>10</u> feet (~~1524~~ <u>3048</u> mm) above the highest fitting connection in that section, or to the highest point in the completed

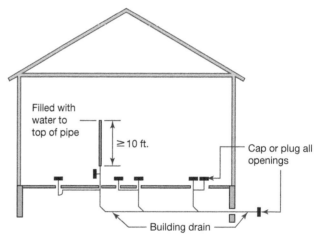

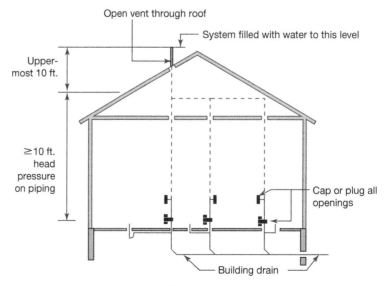

Water test on DWV system.

system. Water shall be held in the section under test for a period of 15 minutes. The system shall prove leak free by visual inspection.

2. Air test. The portion under test shall be maintained at a gauge pressure of 5 pounds per square inch (psi) (34 kPa) or 10 inches of mercury column (34 kPa). This pressure shall be held without introduction of additional air for a period of 15 minutes.

3. <u>Vacuum Test. The portion under test shall be evacuated of air by a vacuum type pump to achieve a uniform gauge pressure of -5 pounds per square inch or a negative 10 inches of mercury column (- 34 kPa). This pressure shall be held without the removal of additional air for a period of 15 minutes.</u>

CHANGE SIGNIFICANCE: Prior to the 2015 edition, the IRC required a 10-foot head pressure for testing drain, waste and vent (DWV) systems with water. In the 2015 code, the head pressure was reduced to 5 feet. Although the 10-foot head pressure has been a long-standing tradition, the change in 2015 was based on an assumption that the actual head pressure is not nearly as critical as the visual nature of the test, and that a 10-foot head test was unlikely to reveal any leaks or defects that would not be detected by a 5-foot head water test.

In the 2021 IRC, the test pressure has been changed back to 10 feet. Reasons for changing back included inconsistency with the required air pressure (when testing with air), which was not decreased in the 2015 IRC. The 5-foot height also does not match the 10-foot height required in the *International Plumbing Code* (IPC).

In this water test, the DWV systems are filled with water to a point 10 feet higher than the piping being tested and the piping and joints are visually inspected for any leaks that might develop. The duration of the water test is 15 minutes to ensure that the system is water tight. The top 10 feet of the DWV systems, which is typically the highest vent through the roof, is only filled with water to the top of the vent terminal. Adding an additional 10-foot standpipe above the vent terminal would not be easily accomplished and would not provide any benefit because the vent will not carry water and not be under pressure in service.

New to the 2021 IRC, a vacuum air test is permitted for testing plastic DWV systems. This alternate test is a means for testing plastic piping systems when the ambient temperatures are below freezing and testing with water presents a challenge. There is no safety hazard in testing with a vacuum such as has occurred in the past with a positive air pressure test on plastic piping. The equipment to perform the test is readily available on the market and a number of contractors are performing the test. Vacuum air testing provides an additional option for testing of plastic piping systems.

P2708.4, P2713.3

Shower and Bathtub Control Valves

CHANGE TYPE: Clarification

CHANGE SUMMARY: The code now addresses field adjustment and access to shower control valves. Lower flow shower heads need to be compatible with the shower control mixing valve.

2021 CODE: P2708.4 Shower control valves. Individual shower and tub/shower combination valves shall be ~~equipped with control valves of the pressure-balance~~ balanced-pressure, thermostatic~~-mixing~~ or combination balanced-pressure~~balance/thermostatic-mixing valve types with a high limit stop in accordance with~~ /thermostatic valves that conform to the requirements of ASSE 1016/ASME A112.1016/CSA B125.16~~. The high limit stop shall be set to limit the water temperature to not greater than 120°F (49°C).~~ or ASME A112.18.1/CSA B125.1. Shower control valves shall be rated for the flow rate of the installed showerhead. Such valves shall be installed at the point of use. Shower and tub/shower combination valves required by this section shall be equipped with a means to limit the maximum setting of the valve to 120°F (49°C), which shall be field adjusted in accordance with the manufacturer's instructions to provide water at a temperature not to exceed 120°F. In-line thermostatic valves shall not be ~~used~~ utilized for compliance with this section.

P2713.3 Bathtub and whirlpool bathtub valves. ~~Hot water supplied to bathtubs~~ Bathtub and whirlpool ~~bathtubs~~ bathtub valves shall ~~be limited to a temperature of not greater than 120°F (49°C)~~ have or be supplied by a water-temperature limiting device that conforms to ASSE 1070/ASME A112.1070/CSA B125.70 ~~or CSA B125.3~~, except where such ~~protection is otherwise provided by a~~ valves are combination tub/shower valve<u>s</u> in accordance with Section P2708.4. The water temperature limiting device required by this section shall be equipped with a means to limit the maximum setting of the device to 120°F (49°C), and, where adjustable,

Shower mixing valve.

shall be field adjusted in accordance with the manufacturer's instructions to provide hot water at a temperature not to exceed 120°F (49°C). Access shall be provided to water temperature limiting devices that conform to ASSE 1070/ASME A112.1070/CSA B125.70.

> **Exception:** Access is not required for non-adjustable water temperature limiting devices that conform to ASSE 1070/ASME A112.1070/CSA B125.70 and are integral with a fixture fitting, provided that the fixture fitting itself can be accessed for replacement.

CHANGE SIGNIFICANCE: The IRC requires a shower head flow rate not greater than 2.5 gallons per minute (gpm). Shower controls (mixing valves) complying with the requirements of the product standards in the code are designed to safety operate at that flow. However, there are many different types of shower heads available in the market that have lower flow rates. These lower flows might not provide enough motive power within a mixing valve to allow for the valve to self-adjust to prevent temperature shock to the user when a supply pressure or temperature change occurs. Not all shower controls (mixing valves) are designed and tested to accommodate lower than 2.5 gpm flows. Section P2708.4 now specifically requires shower control valves to be rated for the flow rate of the installed showerhead to provide the scald and thermal shock protection required by the recognized standard. For bathtubs and whirlpool bathtubs, the code also stipulates that access must be provided to water temperature limiting devices for field adjustment, unless they are nonadjustable water temperature limiting devices.

P2904
Installation Practices for Residential Sprinklers

CHANGE TYPE: Modification

CHANGE SUMMARY: Revisions throughout Section P2904 expand installation practices to more closely align with NFPA 13D.

2021 CODE: P2904.2.1 Temperature rating and separation from heat sources. Except as provided for in Section P2904.2.2, sprinklers shall have a temperature rating of not less than 135°F (57°C) and not more than ~~170°F (77°C)~~ 225°F (107°C). Sprinklers shall be separated from heat sources as required by the sprinkler manufacturer's installation instructions.

P2904.2.3 Freezing areas. Piping shall be protected from freezing as required by Section P2603.5 <u>or by using one of the following:</u>

1. <u>A dry pipe automatic sprinkler system that is listed for residential occupancy applications.</u>
2. ~~Where sprinklers are required in areas that are subject to freezing, dry~~ <u>Dry</u> side-wall or dry-pendent sprinklers extending from a nonfreezing area into a freezing area ~~shall be installed~~.

P2904.3.2 Shutoff valves prohibited. With the exception of shutoff valves for the entire water distribution system <u>or a single master control valve for the automatic sprinkler system that is locked in the open position</u>, valves shall not be installed in any location where the valve would isolate piping serving one or more sprinklers.

P2904.4 Determining system design flow. The flow for sizing the sprinkler piping system shall be based on ~~the flow rating of each sprinkler in accordance with Section P2904.4.1 and the calculation in accordance with Section P2904.4.2.~~ <u>Sections P2904.4.1 and P2904.4.2.</u>

Sprinkler system piping.

P2904.4.1 Determining required flow rate for each sprinkler. The minimum required flow for each sprinkler shall be determined using the sprinkler manufacturer's published data for the specific sprinkler model based on all of the following:

1. The area of coverage.
2. The ceiling configuration, in accordance with Sections P2904.4.1.1 through P2904.4.1.3.
3. The temperature rating.
4. Any additional conditions specified by the sprinkler manufacturer.

P2904.4.1.1 Ceiling configurations. Manufacturer's published flow rates for sprinklers tested under a ceiling 8 feet (2438 mm) in height, in accordance with the sprinkler listing, shall be used for the following ceiling configurations, provided that the ceiling surface does not have significant irregularities, lumps or indentations and is continuous in a single plane.

1. Ceilings that are horizontal or that have a slope not exceeding 8 units vertical in 12 units horizontal (67 percent), without beams, provided that the ceiling height, measured to the highest point, does not exceed 24 feet (7315 mm) above the floor. Where the slope exceeds 2 units vertical in 12 units horizontal (17 percent), the highest sprinkler installed along the sloped portion of a ceiling shall be positioned above all communicating openings connecting the sloped ceiling compartment with an adjacent space.

2. Ceilings that are horizontal or that have a slope not exceeding 8 units vertical in 12 units horizontal (67 percent), with beams, provided that the ceiling height, measured to the highest point, does not exceed 24 feet (7315 mm) above the floor. Beams shall not exceed 14 inches (350 mm) in depth, and pendent sprinklers shall be installed under the beams as described at the end of this section. The compartment containing the beamed ceiling shall not exceed 600 square feet (56 m^2) in area. Where the slope does not exceed 2 units vertical in 12 units horizontal (17 percent), the highest sprinkler in the compartment shall be above all communicating openings connecting the compartment with an adjacent space. Where the slope exceeds 2 units vertical in 12 units horizontal (17 percent) the highest sprinkler installed along the sloped portion of a ceiling shall be positioned above all communicating openings connecting the sloped ceiling compartment with an adjacent space.

3. Ceilings that have a slope exceeding 2 units vertical in 12 units horizontal (17 percent) but not exceeding 8 units vertical in 12 units horizontal (67 percent), with beams of any depth, provided that the ceiling height, measured to the highest point, does not exceed 24 feet (7315 mm) above the floor. Sidewall or pendent sprinklers shall be installed in each pocket formed by beams. The compartment containing the sloped, beamed ceiling shall not exceed 600 square feet (56 m^2) in area.

Pendent, recessed pendent, and flush-type pendent sprinklers installed directly under a beam having a maximum depth of 14 inches (356 mm) shall have the sprinkler deflector located not less than 1 inch (25 mm) or more than 2 inches (51 mm) below the bottom of the beam. Pendent sprinklers installed adjacent to the bottom of a beam having a maximum depth of 14 inches (356 mm) shall be positioned such that the vertical centerline of the sprinkler is no more than 2 inches (51 mm) from the edge of the beam, with the sprinkler deflector located not less than 1 inch (25 mm) or more than 2 inches (51 mm) below the bottom of the beam. Pendent sprinklers shall also be permitted to be installed less than 1 inch (25 mm) below the bottom of a beam where in accordance with manufacturer's instructions for installation of flush sprinklers.

P2904.4.1.3 Other Ceiling Configurations. For ceiling configurations not addressed by Sections P2904.4.1.1 or P2904.4.1.2, the flow rate shall be subject to approval by the building official.

TABLE P2904.6.2(2) Minimum Water Meter Pressure Loss (PL$_m$)[a]

Flow Rate (gallons per minute, gpm)[b]	5/8-Inch Meter Pressure Loss (pounds per square inch, psi)	3/4-Inch Meter Pressure Loss (pounds per square inch, psi)	1-Inch Meter Pressure Loss (pounds per square inch, psi)
8	~~2~~ 3	~~1~~ 3	1
10	3	~~1~~ 3	1
12	4	~~1~~ 3	1
14	~~5~~ 6	~~2~~ 5	1
16	7	~~3~~ 6	1
18	9	~~4~~ 7	~~1~~ 2
20	11	~~4~~ 9	2
~~22~~ 23	~~NP~~ 14	~~5~~ 11	~~2~~ 3
~~24~~	~~NP~~	5	~~2~~
26	~~NP~~ 18	~~6~~ 14	~~2~~ 3
~~28~~	~~NP~~	~~6~~	~~2~~
~~30~~ 31	~~NP~~ 26	~~7~~ 22	~~2~~ 4
~~32~~ 39	~~NP~~ 38	~~7~~ 35	~~3~~ 6
~~34~~	~~NP~~	~~8~~	~~3~~
~~36~~ 52	NP	~~8~~ NP	~~3~~ 10

For SI: 1 inch = 25.4 mm, 1 pound per square inch = 6.895 kPa, 1 gallon per minute = 0.063 L/s. NP = Not permitted unless the actual water meter pressure loss is known.

a. Table P2904.6.2(2) establishes conservative values for water meter pressure loss or installations where the water meter loss is unknown. Where the actual water meter pressure loss is ~~known~~ published and available from the meter manufacturer, PL_m shall be the ~~actual loss.~~ published pressure loss for the selected meter.
b. Flow rate from Section P2904.4.2. Add 5 gpm to the flow rate required by Section P2904.4.2 where the water service pipe supplies more than one dwelling.

Note: Not all changes to Section P2904 are shown. Please refer to the 2021 IRC for the complete text.

CHANGE SIGNIFICANCE: Section P2904 for dwelling unit fire sprinkler systems has been revised to reflect current installation practices and to correlate with some common acceptable methods in NFPA 13D. Section P2904 states that installations must comply with either NFPA 13D or Section P2904, and that Section P2904 is considered equivalent to NFPA 13D.

NFPA 13D allows intermediate temperature sprinklers to be used in lieu of ordinary temperature sprinklers in dwelling units, even where elevated ambient temperatures are not expected. Provided the sprinkler qualifies as a residential sprinkler, there is no reason for activation temperature to be limited in the IRC. Permitting intermediate temperature sprinklers to be used may help ensure that intermediate temperature sprinklers will be present in locations where elevated ambient temperatures are anticipated. Adding this option to the IRC allows more flexibility.

In areas subject to freezing, the code has required dry sidewall or dry-pendent sprinklers extending from a nonfreezing area into a freezing area. The code now permits a dry pipe automatic sprinkler system listed for residential occupancy to be installed as a viable option that is currently permitted in NFPA 13D.

In another correlation change with NFPA 13D, a control valve is now allowed on a standalone sprinkler system. Although NFPA 13D allows such valves to be electronically monitored in lieu of locking, the new language in the IRC requires that the valve be locked open, which could be accomplished by simply providing a nylon strap to secure the valve handle. Although this adds an allowance to have a master control valve, there is no mandate to include one.

Revisions to Section P2904.4.1 on flow rates based on ceiling configuration have been expanded to provide necessary information on sloped and beamed ceiling configurations. A 2010 Fire Protection Research Foundation study helped standardize sprinkler protection criteria for ceilings with slopes and beams by determining that many sloped and beamed ceilings can be suffciently protected using the same design criteria that apply to horizontal ceilings. This prompted NFPA to add model design criteria to their sprinkler standards and manufacturers are now amending cut sheets to delete criteria for ceiling configurations that are covered by NFPA's standards. The IRC now also provides suffcient guidance for protection of sloped or beamed ceilings in sprinklered dwellings.

The water meter table has been revised to better correlate with NFPA 13D. Revised Table P2904.6.2(2) includes NFPA 13D correlated values for 5/8-inch, 3/4-inch and 1-inch meters, and it retains entries for flows less than 18 gpm. NFPA 13D does not include these lesser flows, but the IRC has always included them since low-flow systems are an option for affordable housing. The table does not include values for 1 ½-inch and 2-inch meters, which are included in NFPA 13D because these meter sizes are unnecessary for flow demands associated with home fire sprinkler systems.

The following is a summary of significant changes to Section P2904:

- Section P2904.2.1: Permits intermediate temperature sprinklers to be used in lieu of ordinary temperature sprinklers
- Section P2904.2.3: Allows a listed dry pipe residential sprinkler system for freeze protection
- Section P2904.3.2: Permits a control valve on a standalone sprinkler system
- Section P2904.4.1: Correlates with NFPA 13D and current installation practices for protecting spaces with sloped or beamed ceilings
- Table P2904.6.2(2): Revises the water meter table in the IRC to better correlate with the water meter table in NFPA 13D

P2905.3

Length of Hot Water Piping to Fixtures

CHANGE TYPE: Addition

CHANGE SUMMARY: The code now limits the length of hot water piping serving fixtures.

2021 CODE: <u>**P2905.3 Hot water supply to fixtures.** The developed length of hot water piping, from the source of hot water to the fixtures that require hot water, shall not exceed 100 feet (30 480 mm). Water heaters and recirculating system piping shall be considered to be sources of hot water.</u>

CHANGE SIGNIFICANCE: A new section limits the hot water supply line length to 100 feet measured from the source of hot water to the fixtures that require hot water. This provision is similar to existing language in IPC Section 607.2, except the IPC limits the length to not greater than 50 feet. Hot water supply lines greater than 100 feet waste water (proportional to pipe size) while occupants wait for hot water to reach fixtures for bathing, washing and culinary purposes. Even when hot water supply lines are insulated, the hot water remaining in the lines between demand periods cools down. Limiting the length and consequent volume of heated water in the supply lines reduces the amount of wasted water and occupant waiting time.

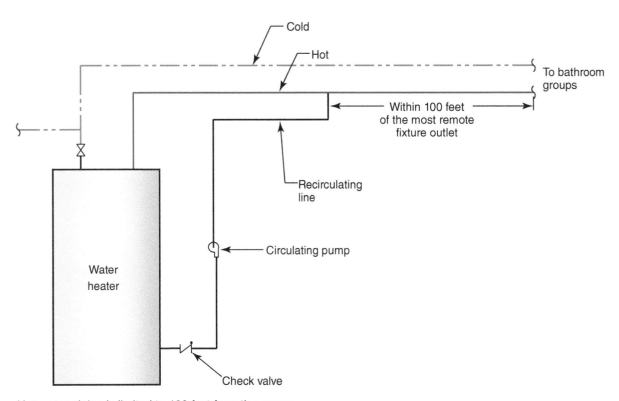

Hot water piping is limited to 100 feet from the source.

P3005.2.10.1
Removable Fixture Traps as Cleanouts

CHANGE TYPE: Clarification

CHANGE SUMMARY: Removable traps and removeable fixtures with integral traps are acceptable for use as cleanouts.

2021 CODE: P3005.2.10.1 Cleanout Equivalent. A fixture trap or a fixture with an integral trap, removable without altering the concealed piping shall be acceptable as a cleanout equivalent.

CHANGE SIGNIFICANCE: Removable traps and removeable fixtures with integral traps such as water closets have long been considered as acceptable access to the drainage system for clearing stoppages. New Section P3005.2.10.1 verifies acceptance of these practices. Note that the existing exception to Section P3005.2.5 (Exception 1) only allows cleanout access through a removable P-trap for the same or one size larger pipe size. Section P3005.2.3 prohibits removal of a water closet to serve as cleanout access to a building sewer.

Removable fixture trap as cleanout equivalent.

P3011

Relining of Building Sewers and Building Drains

CHANGE TYPE: Modification

CHANGE SUMMARY: The code recognizes various available technologies for relining of existing building sewer and building drainage piping and clarifies the inspection procedures.

2021 CODE:

Section P3011
~~Replacement~~ Relining of ~~Underground~~ Building Sewers ~~by PVC Fold and Form Methods~~ Building Drains

P3011.1 General. This section shall govern the ~~replacement~~ relining of existing building sewer ~~piping by PVC Fold and Form methods~~ and building drainage piping.

P3011.2 Applicability. The ~~replacement~~ relining of existing building sewer piping ~~by PVC fold and form methods~~ and building drainage piping shall be limited to gravity drainage piping 4 inches (102 mm) ~~to 18 inches (457 mm). The replacement~~ in diameter and larger. The relined piping shall be of the same nominal size as the existing piping.

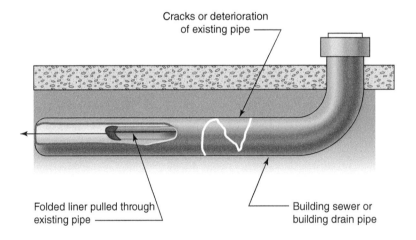

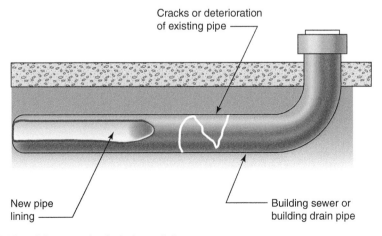

Fold and form method of pipe relining.

P3011.3 Preinstallation ~~inspection~~ requirements. ~~The~~ <u>Prior to commencement of the relining installation, the</u> existing piping sections to be ~~replaced~~ <u>relined shall be descaled and cleaned. After the cleaning process has occurred and water has been flushed through the system, the piping shall be inspected internally by a recorded video camera survey.</u> ~~The survey shall include notations of the position of cleanouts and the depth of connections to the existing piping.~~

<u>**P3011.3.1 Pre-installation recorded video camera survey.** The video survey shall include verification of the project address location. The video shall include notations of the cleanout and fitting locations, and the approximate depth of the existing piping. The video shall also include notations of the length of piping at intervals no greater than 25 feet (7620 mm).</u>

<u>**P3011.4 Permitting.** Prior to issuing a permit for relining, the building official shall review and evaluate the preinstallation recorded video camera survey to determine whether the piping system is able to be relined in accordance with the proposed lining system manufacturer's installation requirements and applicable referenced standards.</u>

~~**P3011.4 Pipe.** The replacement piping shall be manufactured in compliance with ASTM F1504 or ASTM F1871.~~

<u>**P3011.5 Prohibited applications.** Where the preinstallation recorded video camera survey reveals that piping systems are not installed correctly, or defects exist, relining shall not be permitted. The defective portions of piping shall be exposed and repaired with pipe and fittings in accordance with this code. Defects shall include, but are not limited to, backslope or insufficient slope, complete pipe wall deterioration or complete separations such as from tree root invasion or improper support.</u>

~~**P3011.5 Installation.** The piping sections to be replaced shall be cleaned and flushed. Remediation shall be performed where there is groundwater infiltration, roots, collapsed pipe, dropped joints, offsets more than 12 percent of the inside pipe diameter or other obstructions.~~

P3011.6 Relining materials. The relining materials shall be manufactured in compliance with applicable standards and certified as required in Section P2609. <u>Fold-and-form pipe reline materials shall be manufactured in compliance with ASTM F1504 or ASTM F1871.</u>

~~**P3011.6 Cleanouts.** Where the existing building sewer did not have cleanouts meeting the requirements of this code, cleanout fittings shall be installed as required by this code.~~

P3011.7 Installation. The installation of relining materials shall be performed in accordance with the manufacturer's installation instructions, applicable referenced standards and this code.

<u>**P3011.7.1 Material data report.** The installer shall record the data as required by the relining material manufacture and applicable standards. The recorded data shall include but is not limited to the location of the</u>

project, relining material type, amount of product installed and conditions of the installation. A copy of the data report shall be provided to the building official prior to final approval.

P3011.7 Post-installation inspection. ~~The completed replacement piping shall be inspected internally by a recorded video camera survey. The video survey shall be reviewed and approved by the building official prior to pressure testing of the replacement piping system.~~

P3011.8 Post-installation recorded video camera survey. The completed relined piping system shall be inspected internally by a recorded video camera survey after the system has been flushed and flow tested with water. The video survey shall be submitted to the code official prior to finalization of the permit. The video survey shall be reviewed and evaluated to provide verification that no defects exist. Any defects identified shall be repaired and replaced in accordance with this code.

P3011.8 Pressure testing. ~~The replacement piping system as well as the connections to the replacement piping shall be tested in accordance with Section P2503.4.~~

P3011.9 Certification. Certification shall be provided in writing to the building official, from the permit holder, that the relining materials have been installed in accordance with the manufacturer's installation instructions, the applicable standards and this code.

P3011.10 Approval. Upon verification of compliance with the requirements of Sections P3011.1 through P3011.9, the building official shall approve the installation.

CHANGE SIGNIFICANCE: Pipe relining technology reduces or eliminates the need for open trench excavations. The IRC previously recognized relining of underground building sewers by the PVC fold and form method. A factory-made extruded thermoplastic "folded" liner is pulled into a cleaned existing pipe, expanded with air and steam (steam for softening the liner to allow unfolding) and then allowed to cool (using air pressure only) to form a new pipe within the existing pipe. There are other technologies currently available to reline piping systems, which are being approved through the alternate material and methods provisions without guidance from the code. Section 3011 has been expanded to include other approved materials and technologies currently available in the industry. Changes also recognize that pipe relining for restoring existing piping is suitable for building drains in addition to building sewers. The new text provides installation and acceptance criteria for the available methodologies to promote consistent application.

PART 8
Electrical

Chapters 34 through 43

- **Chapter 34** General Requirements
  No changes addressed
- **Chapter 35** Electrical Definitions
  No changes addressed
- **Chapter 36** Services
- **Chapter 37** Branch Circuit and Feeder Requirements
- **Chapter 38** Wiring Methods
  No changes addressed
- **Chapter 39** Power and Lighting Distribution
- **Chapter 40** Devices and Luminaires
  No changes addressed
- **Chapter 41** Appliance Installation
  No changes addressed
- **Chapter 42** Swimming Pools
  No changes addressed
- **Chapter 43** Class 2 Remote-Control, Signaling and Power-Limited Circuits No changes addressed

The electrical part of the IRC is extracted, by permission, from NFPA 70 *National Electrical Code* (NEC) published by the National Fire Protection Association (NFPA). The corresponding NEC section number appears in brackets at the end of each IRC section. Similar to the mechanical, fuel gas, and plumbing parts of the IRC, Part 8 is divided into several chapters, starting with general requirements applicable to all residential electrical systems and followed by chapters of technical provisions covering design and installation. Chapter 34 covers general requirements such as component identification, equipment location, clearances, protection from damage and conductor connections. Chapter 35 of the IRC provides definitions specific to electrical installations and that supplement (and in some cases supersede) the general definitions found in Chapter 2. Subsequent chapters cover electrical services, branch circuits, feeders, wiring methods, outlet locations, receptacles, lighting fixtures and appliance installation for electrical systems of buildings under the scope of the IRC. A separate chapter covers the unique hazards and special requirements related to electrical installations for swimming pools, hot tubs, and whirlpool bathtubs. Limited-voltage circuits are addressed in Chapter 43. ■

E3601.8
Emergency Service Disconnects

E3606.5
Service Surge-Protective Device

E3703.4
Bathroom Branch Circuits

E3703.5
Garage Branch Circuits

E3901.4
Kitchen Countertop and Work Surface Receptacles

E3902
GFCI Protection for 250-Volt Receptacles

E3902.5
GFCI Protection for Basement Receptacles

E3902.10
GFCI Protection for Indoor Damp and Wet Locations

E3601.8 Emergency Service Disconnects

CHANGE TYPE: Addition

CHANGE SUMMARY: An emergency service disconnect is required in a readily accessible outdoor location.

2021 CODE: E3601.8 Emergency disconnects. For one- and two-family dwelling units, all service conductors shall terminate in disconnecting means having a short-circuit current rating equal to or greater than the available fault current, installed in a readily accessible outdoor location. If more than one disconnect is provided, they shall be grouped. Each disconnect shall be one of the following.

1. Service disconnects marked as follows: EMERGENCY DISCONNECT, SERVICE DISCONNECT
2. Meter disconnect switches that have a short-circuit current rating equal to or greater than the available fault current and all metal housings and service enclosures are grounded in accordance with Section E3908.7 and bonded in accordance with Section E3609.

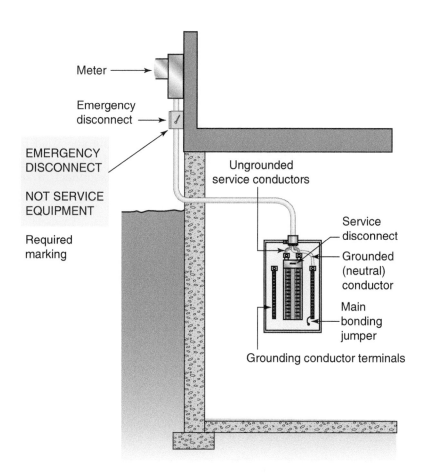

Emergency disconnect required at a readily accessible outdoor location.

A meter disconnect switch shall be capable of interrupting the load served and shall be marked as follows: EMERGENCY DISCONNECT, METER DISCONNECT, NOT SERVICE EQUIPMENT

3. Other listed disconnect switches or circuit breakers on the supply side of each service disconnect that are suitable for use as service equipment and marked as follows: EMERGENCY DISCONNECT, NOT SERVICE EQUIPMENT

Markings shall comply with Section E3404.12. [230.82(3), 230.85]

CHANGE SIGNIFICANCE: The electrical provisions have long required a service disconnecting means either outside of the building or inside the building at a location nearest the point of entrance of the service conductors. Location of the service disconnecting means varies across the United States and is sometimes influenced by geographic and climatic criteria or traditions. In areas with basements, the service disconnect is often in the basement. In other areas, it is inside the garage. In some parts of the country the disconnect is outside the dwelling.

New to the 2021 code, an emergency disconnect is required at a readily accessible location outside the building. The main purpose of the emergency disconnect is to allow first responders to quickly and safely shut down power to the building in an emergency situation. In such a case it is difficult and often unsafe to enter the building to shut down power, particularly if the service disconnect is in the basement. Often first responders may resort to pulling the meter on the outside, a risky endeavor, or they may have to wait for the utility company to disconnect service to the building. Requiring a service disconnect outside intends to resolve these issues.

The outside emergency disconnect may be a separate device mounted outside near the service entrance. Or it may also be the service disconnect, if located outside, or it may be a meter disconnect. In any case a sign or marker is required with text matching the wording specified by the code. The sign or marking must indicate that the device is an emergency disconnect and also whether it serves as a meter disconnect or service disconnect.

E3606.5 Service Surge-Protective Device

CHANGE TYPE: Addition

CHANGE SUMMARY: A surge-protective device (SPD) is now required at the service panel.

2021 CODE: E3606.5 Surge protection. All services supplying one- and two-family dwelling units shall be provided with a surge-protective device (SPD) installed in accordance with Sections E3606.5.1 through E3606.5.3.

E3606.5.1 Location. The SPD shall be an integral part of the service equipment or shall be located immediately adjacent thereto.

Exception: The SPD shall not be required to be located in the service equipment if located at each next-level distribution equipment downstream toward the load.

E3606.5.2 Type. The SPD shall be a Type 1 or Type 2 SPD.

E3606.5.3 Replacement. Where service equipment is replaced, all of the requirements of this section shall apply. [230.67]

CHANGE SIGNIFICANCE: Modern day electronics are prevalent in homes and have become more sensitive to electrical surges, which may damage components or result in data loss. The code now requires a surge-protective device (SPD) located integral to or immediately adjacent to the service equipment. This requirement also applies to replacement service equipment. The SPD must be either a Type 1 or Type 2 device. Type 1 devices are rated for and typically installed on the line (supply) side before the main service disconnect, although they are permitted in Type 2 locations as well. Type 2 devices are only permitted on the load (downstream) side of the main disconnect. The new requirement is in response to an identified need for surge protection of sensitive electronic devices including appliances, GFCI and AFCI devices and smoke alarms.

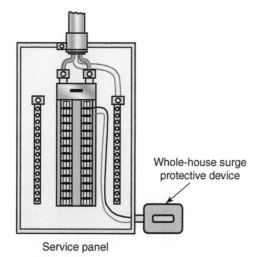

Surge protective device required for service of dwelling unit.

E3703.4
Bathroom Branch Circuits

CHANGE TYPE: Clarification

CHANGE SUMMARY: Only the required bathroom receptacle outlets or those serving a countertop need to be on the dedicated 20-amp bathroom circuit.

2021 CODE: E3703.4 Bathroom branch circuits. A minimum of one 20-ampere branch circuit shall be provided to supply bathroom receptacle outlet(s) <u>required by Section E3901.6 and any countertop or similar work surface receptacle outlets</u>. Such circuits shall have no other outlets. [210.11(C)(3)]

Exception: Where the 20-ampere circuit supplies a single bathroom, outlets for other equipment within the same bathroom shall be permitted to be supplied in accordance with Section E3702. [210.11(C)(3) Exception]

CHANGE SIGNIFICANCE: Section E3703 sets the minimum requirements for branch circuits, including those serving receptacle outlets in specific areas of the dwelling unit. For example, at least two 20-amp circuits are required to serve kitchen countertop receptacle outlets and are permitted to serve other outlets in the kitchen and dining area. For laundry areas and garages, each requires a separate 20-amp circuit for receptacle outlets and these circuits cannot serve other outlets (an exception permits a readily accessible outdoor receptacle outlet to be served by the garage circuit). The rule for bathroom circuits has been a little less well understood. A minimum of one 20-amp branch circuit

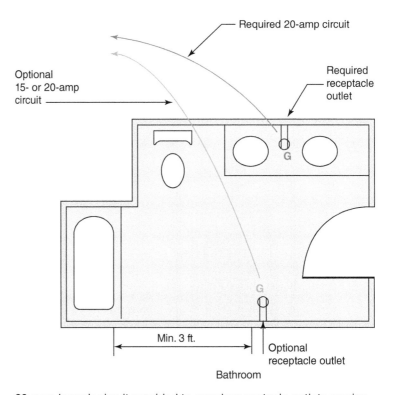

20-amp branch circuit provided to supply receptacle outlets serving bathroom countertop.

is required to serve bathroom receptacle outlets. The code states that this circuit (or circuits) cannot serve other outlets, which would include lighting outlets in the bathrooms, but does permit the circuit to serve multiple bathrooms. However, if the 20-amp circuit supplies only one bathroom, other outlets in that bathroom including lighting outlets may be placed on that circuit. Some have interpreted this provision as requiring any receptacle outlet in the bathroom to be on this 20-amp circuit. That has not been the intent. The 20-amp circuit is intended to serve the required receptacle outlet within 36 inches of the lavatory basin in accordance with Section E3901.6. Other bathroom receptacle outlets are permitted but not required to be on the same circuit. The new text clarifies that only receptacle outlets required by Section E3901.6 and any receptacle outlet serving a countertop or work surface must be on the 20-amp circuit. Other general-purpose receptacle outlets in bathrooms are permitted to be on other 15- or 20-amp circuits.

Note that a new provision in Section E4002.11 prohibits receptacle outlets within 3 feet horizontally from the bathtub rim or shower stall threshold.

E3703.5
Garage Branch Circuits

CHANGE TYPE: Modification

CHANGE SUMMARY: Only the required receptacle outlets must be on the 20-amp dedicated circuit for garages.

2021 CODE: E3703.5 Garage Branch Circuits. In addition to the number of branch circuits required by other parts of this section, not less than one 120-volt, 20-ampere branch circuit shall be installed to supply receptacle outlets <u>required by Section E3901.9</u> in attached garages and in detached garages with electric power. This circuit shall not have other outlets.

Exception: This circuit shall be permitted to supply readily accessible outdoor receptacle outlets.

CHANGE SIGNIFICANCE: Section E3703 sets requirements for the minimum number of branch circuits for a dwelling unit and prescribes the instances requiring separate circuits. For example, a separate branch circuit is required for central heating equipment. Separate 20-amp circuits have been required in three instances – two circuits for kitchen/dining

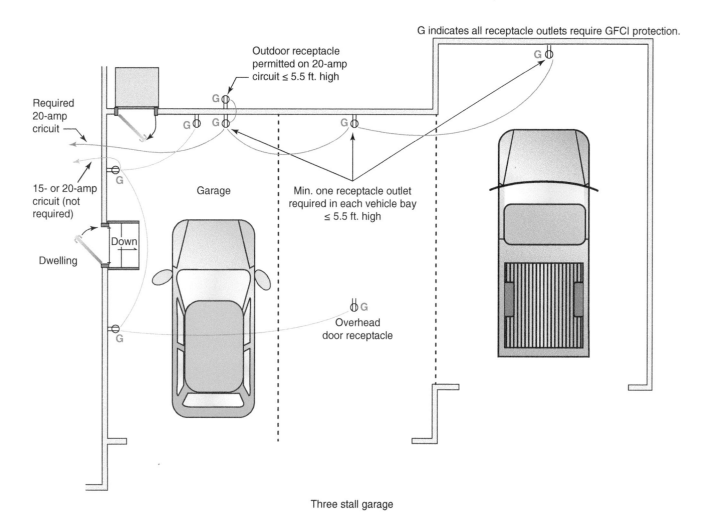

Three stall garage

20-amp branch circuit supplies receptacle outlets required by Section E3901.9.

areas and one each for bathrooms and laundry rooms. The requirement for a separate branch circuit to serve a garage first appeared in the electrical provisions of the 2015 IRC in Section E3901.9. In this case, either a 15- or 20-amp circuit satisfied the code requirement. In the 2018 IRC, the minimum rating for the garage circuit was changed to 20 amps and limited the circuit to serving only the garage receptacle outlets. An exception allows an exterior receptacle outlet on this circuit provided the receptacle has ready access. Lighting outlets in garages are served by 15- or 20-amp general purpose circuits. The new language in the 2021 code specifies that only the required receptacle outlets need to be supplied by this dedicated 20-amp circuit. Section E3109.9 requires a receptacle outlet in each vehicle bay and not greater than 5 feet 6 inches above the floor. Section E3703.5 now permits optional receptacle outlets in the garage, for example the outlet serving the garage door opener, to be on a general purpose 15- or 20-amp circuit.

E3901.4
Kitchen Countertop and Work Surface Receptacles

CHANGE TYPE: Modification

CHANGE SUMMARY: The number of receptacle outlets required for peninsular and island countertops in kitchens is determined by the area of the countertop surface.

2021 CODE: E3901.4 Countertop and work surface receptacles. In kitchens, pantries, breakfast rooms, dining rooms and similar areas of dwelling units, receptacle outlets for countertop and work surfaces <u>that are 12 inches (305 mm) or wider</u> shall be installed in accordance with Sections E3901.4.1 through ~~E3901.4.5~~ <u>E3901.4.3</u> ~~(see Figure E3901.4)~~ <u>and shall not be considered as the receptacle outlets required by Section E3901.2.</u>

<u>For the purposes of this section, where using multi-outlet assemblies containing two or more receptacles, each 12 inches (305 mm) of multi-outlet assembly containing two or more receptacles installed in individual or continuous lengths shall be considered to be one receptacle outlet.</u> (see Figure E3901.4) [210.52(C)]

E3901.4.1 Wall ~~countertop~~ space<u>s</u>. ~~A receptacle outlet shall be installed at each wall countertop and work surface that is 12 inches (305 mm) or wider.~~ Receptacle outlets shall be installed so that no point along the wall line is more than 24 inches (610 mm), measured horizontally, from a receptacle outlet in that space. [210.52(C)(1)]

Exception: Receptacle outlets shall not be required on a wall directly behind a range, counter-mounted cooking unit or sink in the installation described in Figure E3901.4.1. [210.52(C)(1) Exception]

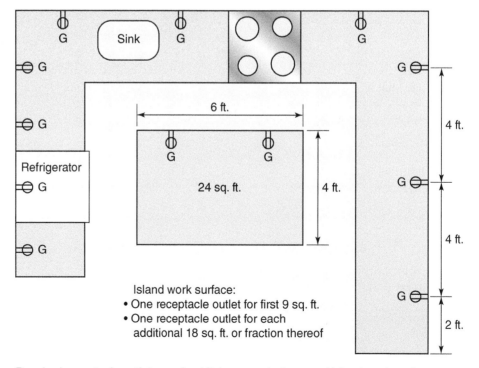

Required receptacle outlets serving kitchen countertops and island work surface.

Significant Changes to the IRC 2021 Edition E3901.4 ■ Kitchen Countertop and Work Surface

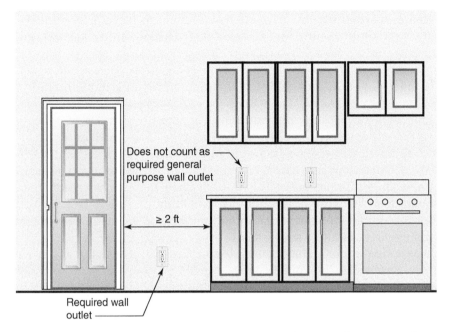

Kitchen countertop receptacle outlet does not count for general purpose wall space outlet.

E3901.4.2 Island and peninsular countertops and work surfaces spaces. Receptacle outlets shall be installed in accordance with the following: [210.52(C)(2)]

1. At least one receptacle outlet shall be ~~installed at each island countertop space with a long dimension of 24 inches (610 mm) or greater and a short dimension of 12 inches (305 mm) or greater.~~ provided for the first 9 square feet (0.84 m^2), or fraction thereof, of the countertop or work surface. A receptacle outlet shall be provided for every additional 18 square feet (1.7 m^2), or fraction thereof, of the countertop or work surface. [210.52(C)(2)(a)]

2. At least one receptacle outlet shall be located within 2 feet (600 mm) of the outer end of a peninsular countertop or work surface. Additional required receptacle outlets shall be permitted to be located as determined by the installer, designer of building owner. The location of the receptacle outlets shall be in accordance with Section E3901.4.3. [210.52(C)(2)(b)]

A peninsular countertop shall be measured from the connected perpendicular wall. [210.52(C)(2)]

3901.4.3 Peninsular countertop space. ~~Not less than one receptacle outlet shall be installed at each peninsular countertop long dimension space having a long dimension of 24 inches (610 mm) or greater and a short dimension of 12 inches (305 mm) or greater. A peninsular countertop is measured from the connected perpendicular wall. [210.52(C)(3)]~~

Note: Not all changes to Section E3901.4 are shown. Please refer to the 2021 IRC for the complete text.

CHANGE SIGNIFICANCE: The kitchen countertop receptacle outlet requirements apply to all of the associated spaces adjoining the kitchen, including dining, breakfast and pantry areas. For kitchen area peninsular and island countertops, the code previously required at least one receptacle outlet for each, provided they met the minimum size requirements of 12 inches by 24 inches. Large kitchens with very large island and peninsular work surfaces are not unusual in modern homes. To adequately serve the various small appliances that may be used on these surfaces, the code now bases the minimum number of outlets on the counter surface area. One receptacle outlet is required for a surface area up to 9 square feet and an additional outlet is required when the area is greater than 9 square feet. For very large counters exceeding 27 square feet, a third outlet is required (an additional outlet for each additional 18 square feet or any fraction of that 18 square feet).

In another minor change, the code clarifies that countertop and work surface receptacles in kitchen areas cannot be counted as a required general-purpose wall space receptacle outlet.

E3902 GFCI Protection for 250-Volt Receptacles

CHANGE TYPE: Modification

CHANGE SUMMARY: Ground-fault circuit-interrupter (GFCI) protection is required for up to 250-volt receptacles in the areas previously identified as requiring GFCI protection for 125-volt receptacles. The 20-amp limitation has been removed.

2021 CODE: E3902.1 Bathroom receptacles. 125-volt through 250-volt, single-phase, 15- and 20-ampere receptacles installed in bathrooms and supplied by single-phase branch circuits rated 150 volts or less to ground shall have ground-fault circuit-interrupter protection for personnel. [210.8(A)(1)]

E3902.2 Garage and accessory building receptacles. 125-volt through 250-volt, single-phase, 15- and 20-ampere receptacles installed in garages and grade-level portions of unfinished accessory buildings used for storage or work areas and supplied by single-phase branch circuits rated 150 volts or less to ground shall have ground-fault circuit-interrupter protection for personnel. [210.8(A)(2)]

E3902.3 Outdoor receptacles. 125-volt through 250-volt, single-phase, 15- and 20-ampere receptacles installed outdoors and supplied by single-phase branch circuits rated 150 volts or less to ground shall have ground-fault circuit-interrupter protection for personnel. [210.8(A)(3)]

Exception: Receptacles as covered in Section E4101.7. [210.8(A)(3) Exception]

E3902.4 Crawl space receptacles and lighting outlets. Where a crawl space is at or below grade level, 125-volt through 250-volt, single-phase, 15- and 20-ampere receptacles installed in such spaces and supplied by single-phase branch circuits rated 150 volts or less to ground

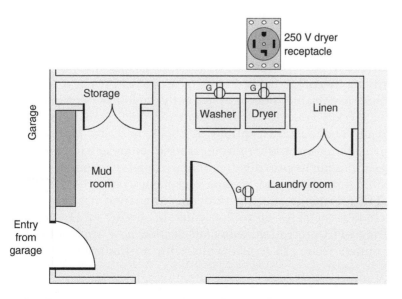

250-volt receptacle outlets requiring GFCI protection.

shall have ground-fault circuit-interrupter protection for personnel. Lighting outlets not exceeding 120 volts shall have ground-fault circuit-interrupter protection. [210.8(A)(4), 2108(E)]

E3902.5 ~~Unfinished b~~Basement receptacles. 125-volt <u>through 250-volt,</u> ~~single-phase, 15- and 20-ampere~~ receptacles installed in ~~unfinished~~ basements <u>and supplied by single-phase branch circuits rated 150 volts or less to ground</u> shall have ground-fault circuit-interrupter protection for personnel. ~~For purposes of this section, unfinished basements are defined as portions or areas of the basement not intended as habitable rooms.~~ [210.8(A)(5)]

> **Exception:** A receptacle supplying only a permanently installed fire alarm or burglar alarm system. Receptacles installed in accordance with this exception shall not be considered as meeting the requirement of Section E3901.9. [210.8(A)(5) Exception]

E3902.6 Kitchen receptacles. 125-volt <u>through 250-volt,</u> ~~single-phase, 15- and 20-ampere~~ receptacles that serve countertop surfaces <u>and supplied by single-phase branch circuits rated 150 volts or less to ground</u> shall have ground-fault circuit-interrupter protection for personnel. [210.8(A)(6)]

E3902.7 Sink receptacles. 125-volt <u>through 250-volt,</u> ~~single-phase, 15- and 20-ampere~~ receptacles that are located within 6 feet (1829 mm) of the top inside edge of the bowl of the sink <u>and supplied by single-phase branch circuits rated 150 volts or less to ground</u> shall have ground-fault circuit-interrupter protection for personnel. [210.8(A)(7)]

E3902.8 Bathtub or shower stall receptacles. 125-volt <u>through 250-volt,</u> ~~single-phase, 15- and 20-ampere~~ receptacles that are located within 6 feet (1829 mm) of the outside edge of a bathtub or shower stall <u>and supplied by single-phase branch circuits rated 150 volts or less to ground</u> shall have ground-fault circuit-interrupter protection for personnel. [210.8(A)(~~8~~<u>9</u>)]

E3902.9 Laundry areas. 125-volt <u>through 250-volt,</u> ~~single-phase, 15- and 20-ampere~~ receptacles installed in laundry areas <u>and supplied by single-phase branch circuits rated 150 volts or less to ground</u> shall have ground-fault circuit-interrupter protection for personnel. [210.8(A)(~~9~~<u>10</u>)]

CHANGE SIGNIFICANCE: GFCI devices protect people from shock hazards by de-energizing a circuit or receptacle when a fault current to ground is detected. For many years the code has required GFCI protection for 125-volt, 15- and 20-amp receptacles in locations where occupants are most susceptible to shock hazards, typically wet or damp locations, areas where water is used, or areas with concrete floor surfaces that might provide a conductive path to ground. The amperage limitation of 15- and 20-amp receptacles has been removed from the GFCI provisions and the 125-volt designation has been expanded to include receptacles up to 250 volts. In the list of 11 specific areas requiring GFCI protection, this code section now applies to all receptacle outlets from 125 to 250 volts with no limitation on amperage.

The change was based on a concern that the higher voltage receptacles in dwelling units also posed a shock hazard to occupants in these specified areas. Typically, the new requirement will apply to 240-volt 30-amp dryer receptacles in laundry rooms. However, any receptacle for a 240-volt appliance in a designated area will require GFCI protection.

E3902.5
GFCI Protection for Basement Receptacles

CHANGE TYPE: Modification

CHANGE SUMMARY: The requirement for GFCI protection in unfinished basement areas has been expanded to include all basement areas.

2021 CODE: E3902.5 Unfinished bBasement receptacles. 125-volt through 250-volt, single-phase, 15- and 20-ampere receptacles installed in unfinished basements and supplied by single-phase branch circuits rated 150 volts or less to ground shall have ground-fault circuit-interrupter protection for personnel. For purposes of this section, unfinished basements are defined as portions or areas of the basement not intended as habitable rooms. [210.8(A)(5)]

> **Exception:** A receptacle supplying only a permanently installed fire alarm or burglar alarm system. Receptacles installed in accordance with this exception shall not be considered as meeting the requirement of Section E3901.9. [210.8(A)(5) Exception]

CHANGE SIGNIFICANCE: The GFCI requirements have been expanded in most 3-year code cycles since 2006. In the 2009 IRC, exceptions related to receptacles serving appliances that were not readily accessible in garages and unfinished basement areas were removed. As a result, all receptacle outlets in unfinished basements (except those for permanently installed fire and burglar alarm systems) and all receptacle outlets in garages have since required GFCI protection. Additional locations requiring GFCI protection appeared in the 2015 edition. Added to the list were 125-volt, single-phase, 15- and 20-amp receptacles located in laundry rooms, which was seen as consistent with GFCI requirements for bathrooms. The kitchen dishwasher branch circuit also first appeared in the 2015 code as requiring GFCI protection. In the 2018 code, lighting outlets in crawl spaces were added to the list.

In the 2021 IRC electrical provisions, GFCI protection is required for all basement areas, whether finished or unfinished. This requirement now applies to habitable space in basements, such as bedrooms, recreation rooms or family rooms and spaces not considered habitable such as

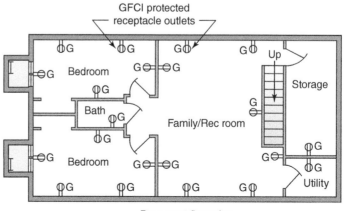

Basement floor plan

All basement receptacles require GFCI protection.

storage or utility areas. The impetus for the change involved concerns about conductive floor surfaces, such as concrete or grouted ceramic tile, in habitable spaces creating potential shock hazards to occupants in areas that may be damp or subject to occasional foundation leakage or flooding. The exemption for receptacles serving fire alarm or burglar alarm system equipment in basements remains in the code. Note that the GFCI provisions now apply to receptacles rated up to 250 volts with no limit on amperage, as discussed in the previous pages for Section E3902.

E3902.10

GFCI Protection for Indoor Damp and Wet Locations

CHANGE TYPE: Addition

CHANGE SUMMARY: GFCI protection is now required for indoor damp and wet locations not included in the other specific locations requiring GFCI protection.

2021 CODE: E3902.10 Indoor damp and wet locations. 125-volt through 250-volt, receptacles installed in indoor damp and wet locations and supplied by single-phase branch circuits rated 150 volts or less to ground shall have ground-fault circuit-interrupter protection for personnel. [210.8(A)(11)]

CHANGE SIGNIFICANCE: The primary concern for shock hazards are those areas with damp or wet surfaces that provide a more conductive path to ground for electricity and are more susceptible to electrical shock incidents. GFCI devices protect people from shock hazards by de-energizing a circuit or receptacle when a fault current to ground is detected. The code prescribes GFCI protection for receptacles in most wet and damp locations related to residential buildings in the IRC. For example, coverage includes outdoor locations, garages, basements, bathrooms and laundry rooms for GFCI protection. The new language is intended to cover any other damp or wet locations that might create similar shock hazards. While this gives another tool to the jurisdiction in protecting the safety of building occupants, this provision will present some interpretation challenges for the building official, who will need to determine its intent and application. The building official has the authority to make such interpretations and to develop consistent policies and procedures under Section R104.1.

As examples, a mud room with access directly from the outdoors or a pet bathing and grooming area might be considered a wet or damp area. Although there may be a sink in these areas and the code already requires GFCI protection for receptacles within 6 feet of a sink, there might be receptacles that are farther than 6 feet from the sink, but still considered part of the wet or damp area. Such factors will need to be evaluated by the building official in determining the level of hazard to occupants of the building.

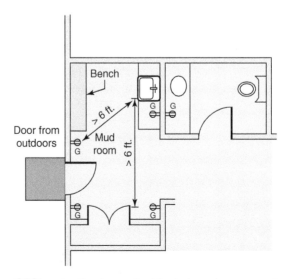

GFCI protection is required for indoor damp or wet locations.

PART 9
Appendices

Appendix A through W

- **Appendices A–E** No changes addressed
- **Appendix F** Radon
- **Appendices G–T** No changes addressed
- **Appendix U** Cob Construction
- **Appendix V** Board of Appeals
     No changes addressed
- **Appendix W** 3D Printed Buildings

As stated in Chapter 1 of the IRC, provisions in the appendices do not apply unless specifically referenced in the adopting ordinance. The appendices are developed in much the same manner as the main body of the model code. However, the appendix information is judged to be outside the scope and purpose of the code at the time of code publication. Many times an appendix offers supplemental information, alternative methods, or recommended procedures. The information may also be specialized and applicable or of interest to only a limited number of jurisdictions. Although an appendix may provide some guidelines or examples of recommended practices or assist in the determination of alternative materials or methods, it will have no legal status and cannot be enforced until it is specifically recognized in the adopting legislation. Appendix chapters or portions of such chapters that gain general acceptance over time can move into the main body of the model code through the code-development process. The 2021 IRC introduces a new Appendix U for cob construction, also called monolithic adobe construction, as well as provisions for 3D printing in Appendix W. ■

AF104
Radon Testing

APPENDIX U
Cob Construction

APPENDIX W
3D Printed Buildings

AF104
Radon Testing

CHANGE TYPE: Addition

CHANGE SUMMARY: Procedures for radon testing are added to Appendix F.

2021 CODE TEXT:

SECTION AF104
TESTING

AF104 Testing. Where radon-resistant construction is required, radon testing shall be as specified in Items 1 through 11:

1. Testing shall be performed after the dwelling passes its air tightness test.
2. Testing shall be performed after the radon control system and HVAC installations are complete. The HVAC system shall be operating during the test. Where the radon system has an installed fan, the dwelling shall be tested with the radon fan operating.

Radon testing apparatus.

3. Testing shall be performed at the lowest occupied floor level, whether or not that space is finished. Spaces that are physically separated and served by different HVAC systems shall be tested separately.
4. Testing shall not be performed in a closet, hallway, stairway, laundry room, furnace room, bathroom or kitchen.
5. Testing shall be performed with a commercially available radon test kit or testing shall be performed by an approved third party with a continuous radon monitor. Testing with test kits shall include two tests, and the test results shall be averaged. Testing shall be in accordance with this section and the testing laboratory kit manufacturer's instructions.
6. Testing shall be performed with the windows closed. Testing shall be performed with the exterior doors closed, except when being used for entrance or exit. Windows and doors shall be closed for at least 12 hours prior to the testing.
7. Testing shall be performed by the builder, a registered design professional, or an approved third party.
8. Testing shall be conducted over a period of not less than 48 hours or not less that the period specified by the testing device manufacturer, whichever is longer.
9. Written radon test results shall be provided by the test lab or testing party. The final written test report with results less than 4 picocuries per liter (pCi/L) shall be provided to the code official.
10. Where the radon test result is 4 pCi/L or greater, the fan for the radon vent pipe shall be installed as specified in Sections AF103.9 and AF103.12.
11. Where the radon test result is 4 pCi/L or greater, the system shall be modified and retested until the test result is less than 4 pCi/L.

Exception: Testing is not required where the occupied space is located above an unenclosed open space.

CHANGE SIGNIFICANCE: Radon is a tasteless colorless radioactive gas that can cause lung cancer. Soil under residences can contain no, low, moderate or high levels of radon. If a jurisdiction decides radon-resistant construction is required, it may adopt *International Residential Code* (IRC) Appendix F. Generally, adoption of Appendix F occurs in Radon Zone 1 and testing of indoor air after the building is completed identifies whether radon levels exceed the established action level. Control systems specified in Appendix F are intended to limit radon entering a residence, but testing is required to determine if further mitigation is needed. Homeowners, potential renters and builders want to know that a radon mitigation system is functional.

The test procedure allows use of either a radon test kit sent to a lab for analysis or a continuous radon monitor. The new test procedures outline when and where testing should occur, whether multiple locations should be tested within a building due to installation of multiple HVAC systems and when retesting is required.

Where radon reduction systems are required, the test procedure serves as the radon system's commissioning process. If a test result is below 4 pCi/L, the level of radon in the building is deemed to be below the action level and further mitigation is not required. Written test results provide a building official and the homeowner with confirmation that a building's radon level is within the prescribed limits. Many buildings pass testing without fan installation, described in Appendix F, by use of a passive radon system. When a passive system does not meet the 4 pCi/L limit, adding a fan typically lowers the radon level below the specified limit.

Significant Changes to the IRC 2021 Edition

Appendix AU
Cob Construction

CHANGE TYPE: Addition

CHANGE SUMMARY: Appendix AU adds a new section on cob construction which has requirements that differ from light straw-clay and strawbale construction.

2021 CODE TEXT:

APPENDIX AU
Cob Construction (Monolithic Adobe)

SECTION AU101
GENERAL

AU101.1 Scope. This appendix provides prescriptive and performance-based requirements for the use of natural cob as a building material. Buildings using cob walls shall comply with this code except as otherwise stated in this appendix.

Cob construction with clay exterior finish.

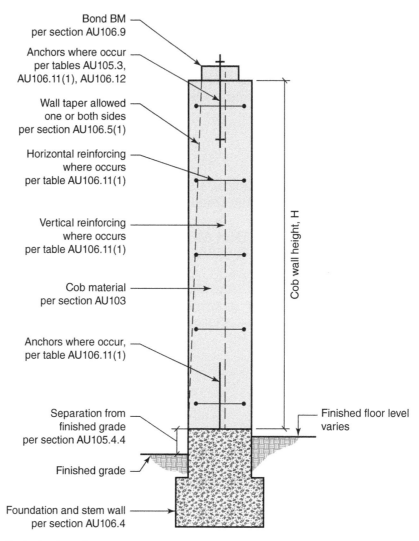

Typical cob wall.

AU101.2 Intent. <u>In addition to the intent described in Section R101.3, the purpose of this appendix is to establish minimum requirements for cob structures that provide flexibility in the application of certain provisions of the code to permit the use of site-sourced and local materials, and to permit combinations of proven historical and modern techniques.</u>

(Only a portion of the appendix is show for brevity and clarity.)

CHANGE SIGNIFICANCE: Cob is an earthen material mix of clay-soil, sand, straw and water, placed onto a wall in layers to create a monolithic wall. Because the material mix and density of cob are very similar to those of adobe bricks, cob is sometimes known as monolithic adobe. Used for thousands of years around the world, the term cob derives from an Old English word for lump, since historical structures were often constructed one handful at a time. Today, cob is often mixed mechanically using a tractor or mortar mixer, but wall construction is still generally manual.

Cob buildings typically feature raised impermeable foundations and extended roof eaves to protect walls from moisture and weather. Walls are often plastered with clay, lime or gypsum plasters that protect the cob without leading to moisture problems associated with less vapor-permeable finishes such as cement stucco, more common on historic adobe structures.

Since the 1990s, there has been interest in cob construction in the United States and much of the world. Cob is highly recyclable, and with good design, construction and maintenance, can withstand centuries of use. Constituent materials are inexpensive compared with lumber, steel, concrete and other commonly used building materials. Cob is noncombustible and nontoxic in all stages of construction and use. The thermal mass and moisture management properties modulate interior temperature and humidity, creating a healthy building.

The cob construction appendix is based on New Zealand's earthen building standards, on U.S. standards for related earthen building systems of adobe and straw-clay and on the experience and testing of cob buildings since the late 1990s.

Appendix AW
3D Printed Buildings

CHANGE TYPE: Addition

CHANGE SUMMARY: Appendix AW adds requirements for 3D printed homes.

2021 CODE TEXT:

<u>**Appendix AW**</u>
<u>**3D PRINTED BUILDING CONSTRUCTION**</u>

<u>**SECTION AW101**</u>
<u>**GENERAL**</u>

<u>**AW101 Scope.** Buildings, structures and building elements fabricated in whole or in part using 3D printed construction techniques shall be designed, constructed and inspected in accordance with the provisions contained in this Appendix and other applicable requirements in this code.</u>

<u>**AW102.1 Definitions.** The words and terms in Section AW102 shall, for the purposes of this appendix, have the meanings shown herein. Refer to Chapter 2 of this code for general definitions.</u>

<u>**SECTION AW102**</u>
<u>**DEFINITIONS**</u>

<u>**3D PRINTED BUILDING CONSTRUCTION.** A process for fabricating buildings, structures and building elements from 3D model data using automated equipment that deposits construction material in a layer upon layer fashion.</u>

3D printing of a home.

ADDITIVE MANUFACTURING MATERIALS. Materials used by the 3D printer to produce the building structure or system components of the building.

FABRICATION PROCESS. Preparation of the job site and construction material, and the deposition, curing, finishing, insertion of components and other methods used to construct building elements such as walls, partitions, roof assemblies and structural components, and the means used to connect assemblies together.

PRODUCTION EQUIPMENT. The equipment, including 3D printer, its settings, nozzles and other accessories used in the fabrication process.

SYSTEM COMPONENTS. Devices, equipment and appliances that are installed in the building elements as part of the wiring, plumbing, HVAC and other systems. These include, but are not limited to, electrical outlet boxes, conduit, wiring, piping, tubing, and HVAC ducts, each of which is covered by a product standard or installation code requirement.

SECTION AW103
BUILDING DESIGN

AW103.1 Design organization. 3D printed buildings, structures and building elements shall be designed by an organization certified in accordance with UL 3401 by an approved agency and approved by the building official in accordance with this section.

AW103.2 Design approval. The structural design, construction documents, and UL 3401 report of findings shall be submitted for review and approval in accordance with Section 104.11 of this code.

SECTION AW104
BUILDING CONSTRUCTION

AW104.1 Construction. 3D printed buildings, structures and building elements shall be constructed in accordance with this section.

AW104.2 Construction method. The building construction method, consisting of the manufacturer's production equipment and fabrication process shall be in accordance with the UL 3401 report of findings. The unique identifier of the construction method used shall match the identifier in the UL 3401 report of findings.

AW104.3 Additive manufacturing materials. Only the listed additive manufacturing materials identified in the UL 3401 report of findings shall be used to fabricate the building structure or system components. Containers of the additive manufacturing materials shall be labeled.

AW104.4 Depositing of manufacturing materials. Manufacturing materials shall only be deposited where ambient temperature and environmental conditions at the job site are within limits specified in the

UL 3401 report of findings. The maximum number of layers permitted, specified curing time and any surface preparation or finishing shall be performed as specified in the UL 3401 report of findings.

SECTION AW105
SPECIAL INSPECTIONS

AW105.1 Initial inspection. An initial inspection of the production equipment, including 3D printer, and the fabrication process shall be performed after the production equipment is located onsite and before building fabrication has begun. The inspection shall be conducted by representatives of the approved agency that evaluated the fabrication process for compliance with UL 3401. The inspection shall verify that the fabrication process, including production equipment, 3D printing parameters and additive manufacturing materials are in accordance with the UL 3401 report of findings, and the proprietary information in the UL 3401 detailed report of findings.

Exception: Where approved by the building official, inspections of the production equipment, including 3D printer, and the fabrication process used in a single housing tract shall be conducted on the first building to be constructed, and on a selected number of subsequent buildings, where the same equipment, equipment operators and fabrication process are used on all buildings. The number of inspections to be performed shall be determined by the building official.

SECTION AW106
REFERENCED STANDARDS

AW106.1 General. See Table AW106.1 for standards that are referenced in various sections of this appendix. Standards are listed by the standard identification with the effective date, standard title, and the section or sections of this appendix that reference the standard.

TABLE AW106.1 Referenced Standards

Standard Acronym	Standard Name	Sections Herein Referenced
UL 3401-19	Outline of Investigation for 3D Printed Building Construction	AW103.2, AW104.2, AW104.3, AW104.4, AW105.1

CHANGE SIGNIFICANCE: 3D building construction has moved from a conceptual stage to reality, and projects are being proposed in an increasing number of jurisdictions. IRC prescriptive design and construction requirements are not applicable to 3D printed fabrication techniques, so code officials have to approve this construction based on limited equivalency evaluations that may not take into account variations in material properties introduced by the 3D printing process, or variation in the physical characteristics of the construction materials used.

UL 3401, *Outline of Investigation for 3D Printed Building Construction*, was developed to evaluate critical aspects of this construction process so that 3D-printed building techniques comply with an equivalent level of safety and performance as legacy construction techniques in the building code.

Appendix AW includes definitions and requirements for 3D-printed building design, construction and special inspections, which rely on designs being evaluated in advance by an approved agency for compliance with UL 3401. Resulting compliance reports include information needed by contractors and code officials to verify applicable code compliance and that the 3D-printing process and on-site materials are the same as those indicated during UL 3401 evaluation and testing. Special inspection is required as portions of the fabrication process such as 3D printer settings, deposition rates and thicknesses and curing processes require special evaluation expertise, particularly when proprietary formulations, equipment and settings are included.

Telework with ICC Digital Codes Premium
An essential online platform for all code users

ICC's Digital Codes is the most trusted and authentic source of model codes and standards, which conveniently provides access to the latest code text from across the United States. With a Premium subscription, users can further enhance their code experience with powerful features such as team collaboration and access sharing, bookmarks, notes, errata and revision history, and more.

Never miss a code update

Available anywhere 24/7

Use on any mobile or digital device

View hundreds of code titles and standards

Go Beyond the Codes with a Premium Subscription
Start Your 14-Day Premium trial at *codes.iccsafe.org/trial*

Proctored Remote Online Testing Option (PRONTO™)

Convenient, Reliable, and Secure Certification Exams

Take the Test at Your Location

Take advantage of ICC PRONTO, an industry leading, secure online exam delivery service. PRONTO allows you to take ICC Certification exams at your convenience in the privacy of your own home, office or other secure location. Plus, you'll know your pass/fail status immediately upon completion.

 With PRONTO, ICC's Proctored Remote Online Testing Option, take your ICC Certification exam from any location with high-speed internet access.

 With online proctoring and exam security features you can be confident in the integrity of the testing process and exam results.,

 Plan your exam for the day and time most convenient for you. PRONTO is available 24/7.

 Eliminate the waiting period and get your results in private immediately upon exam completion.

 ICC was the first model code organization to offer secured online proctored exams—part of our commitment to offering the latest technology-based solutions to help building and code professionals succeed and advance. We continue to expand our catalog of PRONTO exam offerings.

Discover ICC PRONTO and the wealth of certification opportunities available to advance your career: www.iccsafe.org/MeetPRONTO